Student Solutions Manual
for
Physical Chemistry

David W. Ball
Cleveland State University

THOMSON

BROOKS/COLE

Australia • Canada • Mexico • Singapore • Spain • United Kingdom • United States

Printed in the United States of America.
1 2 3 4 5 6 7   06 05 04 03 02

Printed by Globus.
0-534-39714-X

For more information about our products,
contact us at:
Thomson Learning Academic Resource Center
1-800-423-0563

For permission to use material from this text,
contact us by:
Phone: 1-800-730-2214
Fax: 1-800-731-2215
Web: www.thomsonrights.com

Asia
Thomson Learning
5 Shenton Way #01-01
UIC Building
Singapore 068808

Australia
Nelson Thomson Learning
102 Dodds Street
South Street
South Melbourne, Victoria 3205
Australia

Canada
Nelson Thomson Learning
1120 Birchmount Road
Toronto, Ontario M1K 5G4
Canada

Europe/Middle East/South Africa
Thomson Learning
High Holborn House
50-51 Bedford Row
London WC1R 4LR
United Kingdom

Latin America
Thomson Learning
Seneca, 53
Colonia Polanco
11560 Mexico D.F.
Mexico

Spain
Paraninfo Thomson Learning
Calle/Magallanes, 25
28015 Madrid, Spain

PREFACE

This book contains the worked-out end-of-chapter problems to *Physical Chemistry*, 1/e, by David W. Ball. It contains the odd problems only; even problems are worked out in the Instructor Solutions Manual. Also, problems requiring drawings or graphs are not given explicitly here. The reason for that is that there can be a lot of variation in any particular drawing or graph, and rather than present a drawing or graph that is meant to represent the "right" answer, the drawings are left to the student. However, if a numerical or conceptual answer is requested by the drawing exercise, my version of the answer is given. Your version of the answer should be similar, if you did the graph or drawing correctly.

For better or worse, many straightforward questions in physical chemistry require long, involved mathematical manipulations to reach the answer. You, the student, need to accept that up front. Some of the manipulations are algebra, some are calculus: you will be doing a fair share of math in this course. If you are not comfortable with the level of math needed in this course, a refresher of algebra and calculus may be in order. But have no doubt that you will need a proper understanding of algebra and calculus tools in order to master this material. Do not be afraid to ask your instructor for help with the math, in addition to help with the chemistry.

One of the fundamental ideas behind *Physical Chemistry* was to prepare a usable, student-approachable textbook, rather than an encyclopedia from which your instructor picks and chooses. The end-of-chapter exercises were developed in the same vein. While there are over 1000 problems, for the most part each one focuses on a different concept or aspect of the material. Thus, in order to really master the topic, you might want to consider, ultimately, working every problem in each chapter. I've tried to avoid complex questions that require the student to figure out what's being asked first, before trying to solve. The solutions can be challenging enough; there's no need to write a convoluted question!

And watch your units! Many students (and instructors, it seems) don't understand that the units of quantities need to work out algebraically, along with the numbers. This text places special emphasis on units. Indeed, some examples and exercises consider only the units of an expression.

Although every effort has been taken to avoid errors, some may have crept in anyway. If you find any, please let me know, either by letter or email, and I will be happy to credit you if you find it first. I will maintain an Errata file on my personal web site for you to consult. The current web address is academic.csuohio.edu/ball; however, remember that the Web is always changing. If for some reason my web address changes, you should be able find it through my web page at Cleveland State University.

I would like to express my personal thanks to my colleague Mark Waner, a chemistry professor at John Carroll University, who performed an accuracy check of the draft of this manual.

Good luck, and good chemistry to you all!

David W. Ball
Cleveland State University
d.ball@csuohio.edu

CHAPTER 1. GASES AND THE ZEROTH LAW OF THERMODYNAMICS

1.1. The drawing is left to the student. The calorimeter, water bath, and associated equipment (thermometers, ignition system, and so forth) are the system, while the surroundings are everything outside the apparatus.

1.3. (a) $12.56 \, L \times \dfrac{1000 \, mL}{1 \, L} \times \dfrac{1 \, cm^3}{1 \, mL} = 12,560 \, cm^3 = 1.256 \times 10^4 \, cm^3$

(b) $45°C + 273.15 = 318 \, K$

(c) $1.055 \, atm \times \dfrac{1.01325 \, bar}{1 \, atm} \times \dfrac{100,000 \, Pa}{1 \, bar} = 1.069 \times 10^5 \, Pa$

(d) $1233 \, mmHg \times \dfrac{1 \, torr}{1 \, mmHg} \times \dfrac{1 \, atm}{760 \, torr} \times \dfrac{1.01325 \, bar}{1 \, atm} = 1.644 \, bar$

(e) $125 \, mL \times \dfrac{1 \, cm^3}{1 \, mL} = 125 \, cm^3$

(f) $4.2 \, K - 273.15 = -269.0°C$

(g) $25750 \, Pa \times \dfrac{1 \, bar}{100,000 \, Pa} = 0.2575 \, bar$

1.5. In terms of the zeroth law of thermodynamics, heat will flow from the (hot) burner or flame on the stove into the (cold) water, which gets hotter. Then heat will move from the hot water into the (colder) egg.

1.7. For this sample of gas under these conditions, $F(p) = \dfrac{0.0250 \, L}{(-33.0 + 273.15)K} = 1.04 \times 10^{-4} \, L/K$.

(Note the conversion to Kelvin for the temperature.) If the volume is going to be 66.9 cm^3 = 0.0669 L: $0.0669 \, L \times \dfrac{1 \, K}{1.04 \times 10^{-4} \, L} = 643 \, K$.

1.9. There are many possible conversions. Using the fact that 1 cal = 4.184 J:

$8.314 \dfrac{J}{mol \cdot K} \times \dfrac{1 \, cal}{4.184 \, J} = 1.987 \dfrac{cal}{mol \cdot K}$

1.11. Calculations using STP and SATP use different numerical values of R because the sets of conditions are defined using different units. It's still the same R, but it's expressed in different units of pressure, atm for STP and bar for SATP.

1.13. The partial pressure of $N_2 = 0.80 \times \dfrac{14.7 \text{ lb}}{\text{in.}} = 11.8 \dfrac{\text{lb}}{\text{in.}}$

The partial pressure of $O_2 = 0.20 \times \dfrac{14.7 \text{ lb}}{\text{in.}} = 2.9 \dfrac{\text{lb}}{\text{in.}}$

1.15. (a) This is an equation for a straight line with slope = 5, so at $x = 5$ and $x = 10$, the slope is simply 5.

(b) The slope of this function is given by its first derivative: $\dfrac{d}{dx}(3x^2 - 5x + 2) = 6x - 5$. At $x = 5$, the slope is $6(5) - 5 = 25$. At $x = 10$, the slope is $6(10) - 5 = 55$.

(c) The slope of this function is given by its first derivative: $\dfrac{d}{dx}\left(\dfrac{7}{x}\right) = -\left(\dfrac{7}{x^2}\right)$. At $x = 5$, the slope is $-\dfrac{7}{25}$; at $x = 10$, the slope is $-\dfrac{7}{100}$.

1.17. (a) $\left(\dfrac{\partial V}{\partial p}\right)_{T,n} = \dfrac{\partial}{\partial p}\left(\dfrac{nRT}{p}\right) = -\dfrac{nRT}{p^2}$ **(b)** $\left(\dfrac{\partial V}{\partial n}\right)_{T,p} = \dfrac{\partial}{\partial n}\left(\dfrac{nRT}{p}\right) = \dfrac{RT}{p}$ **(c)**

$\left(\dfrac{\partial T}{\partial V}\right)_{p,n} = \dfrac{\partial}{\partial V}\left(\dfrac{pV}{nR}\right) = \dfrac{p}{nR}$ **(d)** $\left(\dfrac{\partial p}{\partial T}\right)_{V,n} = \dfrac{\partial}{\partial T}\left(\dfrac{nRT}{V}\right) = \dfrac{nR}{V}$ **(e)** $\left(\dfrac{\partial p}{\partial n}\right)_{T,V} = \dfrac{\partial}{\partial n}\left(\dfrac{nRT}{V}\right) = \dfrac{RT}{V}$

1.19. $\left(\dfrac{d}{dV}\left(\dfrac{dT}{dp}\right)_V\right)_p$ or $\left(\dfrac{d}{dp}\left(\dfrac{dT}{dV}\right)_p\right)_V$.

1.21. Using $T_B = \dfrac{a}{bR}$, and using data from Table 1.6, we have:

for CO_2: $T_B = \dfrac{3.592 \dfrac{L^2 \text{atm}}{\text{mol}^2}}{(0.04267\ L/\text{mol})\left(0.08205 \dfrac{L\ \text{atm}}{\text{mol K}}\right)} = 1026\ K$

for O_2: $T_B = \dfrac{1.360 \dfrac{L^2 \text{atm}}{\text{mol}^2}}{(0.03183\ L/\text{mol})\left(0.08205 \dfrac{L\ \text{atm}}{\text{mol K}}\right)} = 521\ K$

for N₂:
$$T_B = \frac{1.390 \dfrac{L^2\,atm}{mol^2}}{(0.03913\ L/mol)\left(0.08205\dfrac{L\,atm}{mol\,K}\right)} = 433\ K$$

1.23. The C term is $\dfrac{C}{\overline{V}^2}$. In order for the term to be unitless, C should have units of (volume)^2, or L^2. The C' term is $C'p^2$, and in order for this term to have the same units as pV (which would be L·atm), C' would need units of $\dfrac{L}{atm}$. (The unit bar could also be substituted for atm if bar units are used for pressure.)

1.25. Gases that have lower Boyle temperatures will be most ideal (at least at high temperatures). Therefore, they should be ordered as He, H_2, Ne, N_2, O_2, Ar, CH_4, and CO_2.

1.27. Let us assume standard conditions of temperature and pressure, so $T = 273.15$ K and $p = 1.00$ atm. Also, let us assume a molar volume of $22.412\ L = 2.2412 \times 10^4\ cm^3$. Therefore, we have for hydrogen:
$$\frac{pV}{RT} = 1 + \frac{B}{\overline{V}} = 1 + \frac{15\ cm^3/mol}{2.2412 \times 10^4\ cm^3/mol} = 1.00067,\ \text{which is a 0.067\% increase in the}$$
compressibility. For H_2O, we have:
$$\frac{pV}{RT} = 1 + \frac{B}{\overline{V}} = 1 + \frac{-1126\ cm^3/mol}{2.2412 \times 10^4\ cm^3/mol} = 0.9497,\ \text{which is a 5.0\% decrease in the}$$
compressibility with respect to an ideal gas.

1.29. By comparing the two expressions from the text
$$Z = 1 + \left(b - \frac{a}{RT}\right)\frac{1}{\overline{V}} + \left(\frac{b}{\overline{V}}\right)^2 + \cdots \text{ and } Z = 1 + \frac{B}{\overline{V}} + \frac{C}{\overline{V}^2} + \cdots$$
it seems straightforward to suggest that, at the first approximation, $C = b^2$. Additional terms involving $\overline{V}^2$ may occur in later terms of the first expression, necessitating additional corrections to this approximation for C.

1.31. In terms of p, V, and T, we can also write the following two expressions using the cyclic rule:
$$\left(\frac{\partial V}{\partial T}\right)_p = -\frac{\left(\dfrac{\partial p}{\partial T}\right)_V}{\left(\dfrac{\partial p}{\partial V}\right)_T} \text{ and } \left(\frac{\partial V}{\partial p}\right)_T = -\frac{\left(\dfrac{\partial T}{\partial p}\right)_V}{\left(\dfrac{\partial T}{\partial V}\right)_p}.\ \text{There are other constructions possible that would}$$
be reciprocals of these relationships or the one given in Figure 1.11.

1.33. Since the expansion coefficient is defined as $\frac{1}{V}\left(\frac{\partial V}{\partial T}\right)_p$, α will have units of

$$\frac{1}{\text{volume}} \cdot \frac{\text{volume}}{\text{temperature}} = \frac{1}{\text{temperature}},$$ so it will have units of K^{-1}. Similarly, the isothermal

compressibility is defined as $-\frac{1}{V}\left(\frac{\partial V}{\partial p}\right)_T$, so κ will have units of $\frac{1}{\text{volume}} \cdot \frac{\text{volume}}{\text{pressure}} = \frac{1}{\text{pressure}}$,

or atm^{-1} or bar^{-1}.

1.35. For an ideal gas, $\kappa = -\frac{1}{V}\left(\frac{\partial V}{\partial p}\right)_T = -\frac{1}{V}\frac{\partial}{\partial p}\left(\frac{nRT}{p}\right) = \frac{1}{V}\frac{nRT}{p^2}$. Since $\frac{nRT}{p} = V$, this last

expression becomes $\frac{1}{V}\frac{V}{p} = \frac{1}{p}$ for an ideal gas. The expression $\frac{T}{p}\alpha$ is evaluated as

$$\frac{T}{p}\alpha = \frac{T}{p}\frac{1}{V}\left(\frac{\partial V}{\partial T}\right)_p = \frac{T}{pV}\frac{\partial}{\partial T}\left(\frac{nRT}{p}\right) = \frac{T}{pV}\frac{nR}{p}.$$ For an ideal gas, the ideal gas law can be

rearranged to give $\frac{nR}{p} = \frac{V}{T}$, so we substitute to get that this last expression is

$\frac{T}{pV}\frac{V}{T}$, which $= \frac{1}{p}$. Thus, the two sides of the equation ultimately yield the same expression and

so are equal.

1.37. For an ideal gas, $\overline{V} = \frac{RT}{p}$. Therefore, the expression for density becomes, substituting for

the molar volume, $d = \frac{M}{RT/p} = \frac{pM}{RT}$. The derivative of this expression with respect to

temperature is $\left(\frac{\partial d}{\partial T}\right)_{p,n} = -\frac{pM}{RT^2}$. Using the definition of $\overline{V}$, this can be rewritten as

$$\left(\frac{\partial d}{\partial T}\right)_{p,n} = -\frac{M}{\overline{V}T}.$$

4

CHAPTER 2. THE FIRST LAW OF THERMODYNAMICS

2.1. $\text{work} = \vec{F} \cdot \vec{s} = |F||s|\cos\theta$

(a) $\text{work} = 30 \text{ N} \cdot 30 \text{ m} \cdot \cos 0° = 900 \text{ N·m} = 900 \text{ J}.$

(b) $\text{work} = 30 \text{ N} \cdot 30 \text{ m} \cdot \cos 45° = 900 \cdot 0.7071 \text{ N·m} = 640 \text{ J}.$

2.3. $w = -p_{ext}\Delta V = -(2.33 \text{ atm})(450. \text{ mL} - 50. \text{ mL}) \times \dfrac{1 \text{ L}}{1000 \text{ mL}} = 0.932 \text{ L} \cdot \text{atm} \times \dfrac{101.32 \text{ J}}{1 \text{ L} \cdot \text{atm}} = 94.4 \text{ J}.$

2.5. (a) The work would be less because the external pressure is less.
(b) The work would be greater because the external pressure is greater.
(c) No work would be performed because the external pressure is (effectively) zero.
(d) The work would be greater if the process were irreversible.

2.7. These three compounds experience hydrogen bonding between their molecules. Because it requires more energy to overcome the effects of this hydrogen bonding, these compounds have higher specific heat capacities than other, similar-mass molecules.

2.9. First, calculate the energy needed to warm the water:
$q = m \cdot c \cdot \Delta T = (1.00 \times 10^5 \text{ g})(4.18 \text{ J/g·K})(1.00 \text{ K}) = 4.18 \times 10^5 \text{ J}$

Now, determine how many drops of a 20.0 kg weight falling 2.00 meters in gravity will yield that much energy. The amount of energy in one drop is
$mgh = (20.0 \text{ kg})(9.81 \text{ m/s}^2)(2.00 \text{ m}) = 392.4 \text{ J}.$ Therefore,
$\# \text{drops} = \dfrac{4.18 \times 10^5 \text{ J}}{392.4 \text{ J/drop}} = 1070 \text{ drops}.$

2.11. The verification of equation 2.8 follows from Boyle's law, which says $p_iV_i = p_fV_f$. This can be rearranged to give $\dfrac{V_f}{V_i} = \dfrac{p_i}{p_f}$. Substituting this into equation 2.7:

$w_{rev} = -nRT \ln\dfrac{V_f}{V_i}$ becomes $w_{rev} = -nRT \ln\dfrac{p_i}{p_f}$ (which is what we are supposed to verify).

2.13. Equation 2.10 is not a contradiction of equation 2.11 because equation 2.11 is applied for systems in which the total energy *does* change. This can happen for open or closed systems. Equation 2.10 only applies to *isolated* systems.

2.15. $w = -nRT \ln\dfrac{V_f}{V_i} = -(0.245 \text{ mol})(8.314 \text{ J/mol} \cdot \text{K})(95.0 + 273.15 \text{ K}) \ln\dfrac{0.001 \text{ L}}{1.000 \text{ L}}$

$w = 5180 \text{ J}$

2.17. If any change in a system is isothermal, then the change in U must be zero. It doesn't matter if the process is adiabatic or not!

2.19. Temperature is a state function because an overall change in temperature is determined solely by the initial temperature and the final temperature, not the path a series of temperature changes took.

2.21. First, we should determine the number of moles of gas in the cylinder. Assuming the ideal gas law holds:

$pV = nRT$ can be rearranged to $n = \dfrac{pV}{RT} = \dfrac{(172 \text{ atm})(80.0 \text{ L})}{(0.08205 \dfrac{\text{L} \cdot \text{atm}}{\text{mol} \cdot \text{K}})(20.0 + 273.15 \text{ K})}$

$n = 572 \text{ mol N}_2$ gas

(a) The final pressure can be determined using Charles' law: $\dfrac{p_i}{T_i} = \dfrac{p_f}{T_f}$

$p_f = \dfrac{p_i T_f}{T_i} = \dfrac{(172 \text{ atm})(140.0 + 273.15 \text{ K})}{(20.0 + 273.15 \text{ K})}$ $\quad p_f = 242 \text{ atm}$

(b) $w = 0$ since the volume of the tank does not change.
$q = n \cdot c \cdot \Delta T = (572 \text{ mol})(21.0 \text{ J/mol·K})(140.0°C - 20.0°C) = 1.44 \times 10^6$ J
$\Delta U = q + w = 1.44 \times 10^6$ J $+ 0 = 1.44 \times 10^6$ J.

2.23. $w = -nRT \ln \dfrac{V_f}{V_i} = -(0.505 \text{ mol})(8.314 \dfrac{\text{J}}{\text{mol} \cdot \text{K}})(5.0 + 273.15 \text{ K}) \ln \dfrac{0.10 \text{ L}}{1.0 \text{ L}} = +2690$ J

$q = -1270$ J (given)

$\Delta U = q + w = +2689$ J $- 1270$ J $= +1420$ J

$\Delta H = \Delta U + \Delta(pV)$ Since the process occurs at constant temperature, Boyle's law applies and $\Delta(pV) = 0$. Therefore, $\Delta H = +1420$ J.

2.25. In terms of pressure and volume: $dU = \left(\dfrac{\partial U}{\partial p}\right)_V dp + \left(\dfrac{\partial U}{\partial V}\right)_p dV$.

For enthalpy: $dH = \left(\dfrac{\partial H}{\partial p}\right)_V dp + \left(\dfrac{\partial H}{\partial V}\right)_p dV$.

2.27. In order for each term to have units of J/mol·K for each term, the first term has units J/mol·K; the second term has units J/mol·K^2; and the third term has units J·K/mol.

2.29. $\left(\dfrac{\partial H}{\partial p}\right)_T = \left(\dfrac{\partial (U + pV)}{\partial p}\right)_T = \left(\dfrac{\partial U}{\partial p}\right)_T + \left(\dfrac{\partial pV}{\partial p}\right)_T = 0 + \left(\dfrac{\partial pV}{\partial p}\right)_T = \left(\dfrac{\partial pV}{\partial p}\right)_T = \left(\dfrac{\partial RT}{\partial p}\right)_T$

which is zero at constant temperature. Therefore, $\left(\dfrac{\partial H}{\partial p}\right)_T$ is zero.

2.31. This derivation is given explicitly in the text in section 2.7.

2.33. For He:

$T = \dfrac{2a}{Rb} = \dfrac{2(0.03508\,\mathrm{L}^2 \cdot \mathrm{atm/mol}^2)}{(0.08205\,\mathrm{L} \cdot \mathrm{atm/mol} \cdot \mathrm{K})(0.0237\,\mathrm{L/mol})} = 36.1\,\mathrm{K}$ (compared to 40 K from text)

For H_2:

$T = \dfrac{2a}{Rb} = \dfrac{2(0.244\,\mathrm{L}^2 \cdot \mathrm{atm/mol}^2)}{(0.08205\,\mathrm{L} \cdot \mathrm{atm/mol} \cdot \mathrm{K})(0.0266\,\mathrm{L/mol})} = 224\,\mathrm{K}$ (compared to 202 K from text)

2.35. Because the pressure change isn't too drastic, our answer to exercise 2.34 is probably within a few degrees of being correct – if a truly isenthalpic process can be arranged.

2.37. Because strictly speaking, heat capacities are extensive properties; they depend on the amount of matter in the system. Thus, the form in equation 2.37 is the most general expression that relates the two quantities.

2.39. First, calculate the initial pressure and assume that this is the external pressure that remains constant throughout the compression (a good approximation, since the gas contracts slowly):

$p = \dfrac{nRT}{V} = \dfrac{(0.145\,\mathrm{mol})(0.08205\,\mathrm{L} \cdot \mathrm{atm/mol} \cdot \mathrm{K})(273.15\,\mathrm{K})}{5.00\,\mathrm{L}} = 0.650\,\mathrm{atm}$

Now we can calculate work as $-p_{ext}\Delta V$:

$w = -(0.650\,\mathrm{atm})(3.92\,\mathrm{L} - 5.00\,\mathrm{L}) \times \dfrac{101.32\,\mathrm{J}}{1\,\mathrm{L} \cdot \mathrm{atm}} = +71.1\,\mathrm{J}$

To determine ΔU, we need to calculate q first. We need the final temperature, which can be determined by Charles' law (since pressure is constant):

$\dfrac{V_i}{T_i} = \dfrac{V_f}{T_f} \qquad \dfrac{5.00\,\mathrm{L}}{273.15\,\mathrm{K}} = \dfrac{3.92\,\mathrm{L}}{T_f} \qquad T_f = 214\,\mathrm{K} \qquad$ This means that $\Delta T = (214 - 273.15) = -59\,\mathrm{K}$.

Using $q = n \cdot c \cdot \Delta T$: $q = (0.145\,\mathrm{mol})(20.79\,\mathrm{J/mol \cdot K})(-59\,\mathrm{K}) = -178\,\mathrm{J}$.

$\Delta U = q + w = -178\,\mathrm{J} + 71.1\,\mathrm{J} = -107\,\mathrm{J}$.

2.41. Starting with $-R \ln V \big|^{V_f}_{V_i} = \overline{C}_V \ln T \big|^{T_f}_{T_i}$: both logarithm terms can be evaluated similarly. The logarithm on the left is evaluated as $-R(\ln(V_f) - \ln(V_i))$. Since $\ln(a) - \ln(b) = \ln(a/b)$, this simplifies to $-R \cdot \ln(V_f / V_i)$, which is the left side of equation 2.44. The right side gets evaluated and simplified similarly.

2.43. For an ideal diatomic gas, $\overline{C}_v = \frac{5}{2}R$ and $\overline{C}_p = \frac{7}{2}R$. Performing a similar substitution as in exercise 2.42:

$$\gamma = \frac{\overline{C}_p - \overline{C}_v}{\overline{C}_v} = \frac{\frac{7}{2}R - \frac{5}{2}R}{\frac{5}{2}R} = \frac{\frac{2}{2}R}{\frac{5}{2}R} = \frac{2/2}{5/2} = \frac{2}{5} \text{ for an ideal diatomic gas.}$$

2.45. If the melting process occurs at standard pressure, then $\Delta H = q_p = 333.5$ J (from Table 2.3). To correct for the volume change, we need to calculate the volumes of both water and ice at 0°C:

for water : $1 \text{ gram} \times \dfrac{1 \text{ mL}}{0.99984 \text{ g}} = 1.00016 \text{ mL}$

for ice : $1 \text{ gram} \times \dfrac{1 \text{ mL}}{0.9168 \text{ g}} = 1.0907 \text{ mL}$ Therefore, $\Delta V = 1.00016 - 1.0907 = -0.0905$ mL

Using the equation $\Delta U = \Delta H + \Delta(pV) = \Delta H + p\Delta V$ (for constant pressure):

$$\Delta U = 333.5 \text{ J} + (1 \text{ atm})(-0.0905 \text{ mL})\left(\frac{1 \text{ L}}{1000 \text{ mL}}\right)\left(\frac{101.32 \text{ J}}{1 \text{ L} \cdot \text{atm}}\right) = 333.5 - 0.0090 \text{ J} \approx 333.5 \text{ J}$$

This shows that ΔU and ΔH can be very close, if not virtually the same, for many condensed-phase processes.

2.47. Steam burns hurt more than hot water burns because steam gives up a considerable amount of heat as the heat of vaporization.

2.49. The heat of fusion given up by the freezing water can be transferred (at least in part) to the citrus fruit, keeping them warmer and (hopefully) keeping the fruit itself from freezing.

2.51. $\Delta_{rxn} H = \sum \Delta_f H(\text{prods}) - \sum \Delta_f H(\text{rcts}) = (2 \text{ mol})(26.5 \text{ kJ/mol}) - 0 - 0 = 53.0 \text{ kJ}$

2.53. The reactions are:

$2 \times [\text{NaHCO}_3 \text{ (s)} \rightarrow \text{Na (s)} + \frac{1}{2} \text{H}_2 \text{ (g)} + \text{C (s)} + 3/2 \text{ O}_2 \text{ (g)}]$	$-2 \times \Delta_f H = +1901.62 \text{ kJ}$
$2 \text{ Na (s)} + \text{C (s)} + 3/2 \text{ O}_2 \text{ (g)} \rightarrow \text{Na}_2\text{CO}_3 \text{ (s)}$	$\Delta_f H = -1130.77 \text{ kJ}$
$\text{C (s)} + \text{O}_2 \text{ (g)} \rightarrow \text{CO}_2 \text{ (g)}$	$\Delta_f H = -393.51 \text{ kJ}$
$\text{H}_2 \text{ (g)} + \frac{1}{2} \text{O}_2 \text{ (g)} \rightarrow \text{H}_2\text{O (l)}$	$\Delta_f H = -285.83 \text{ kJ}$

This yields the overall reaction (you can verify that), and the overall $\Delta_{rxn} H$ is the sum of the values on the right: $\Delta_{rxn} H = 91.51$ kJ.

2.55. Since process is constant-volume, $q_V = \Delta U = -31{,}723$ J.

$w = 0$ since the process is in a constant-volume calorimeter. To determine ΔH, we need to know the balanced chemical equation for the combustion of benzoic acid:

$$C_6H_5COOH \text{ (s)} + 15/2\ O_2 \text{ (g)} \rightarrow 7\ CO_2 \text{ (g)} + 3\ H_2O \text{ (l)}$$

For every mole of benzoic acid combusted, there is a change of $(7 - 15/2) = -0.5$ moles of gas. 1.20 grams of benzoic acid are $1.20\ g \times \dfrac{1\ mol}{122.0\ g} = 0.00984\ mol$ of benzoic acid. Therefore, the net change in number of moles of gas is $0.00984 \times (-0.5) = -0.00492$ mol of gas. Using the relationship

$\Delta H = \Delta U + \Delta(pV) = \Delta U + \Delta(nRT) = \Delta U + (\Delta n)RT$, we can determine the ΔH of the process:

$\Delta H = -31{,}723\ J + (-0.00492\ mol)(8.314\ J/mol{\cdot}K)(24.6 + 273.15\ K) = -31{,}723 - 12.2 = -31{,}735\ J$.

2.57. This problem is very similar to Example 2.19, so we will follow that example, taking data from Table 2.1.

The heat needed to bring the reactants from 500°C (or 773 K) to 298 K is:
$\Delta H_1 = q = (2\ mol)(2\ g/mol)(14.304\ J/g{\cdot}K)(-475\ K) + (1\ mol)(32\ g/mol)(0.918\ J/g{\cdot}K)(-475\ K)$
$= -41131\ J$

The heat of reaction is $2 \times (\Delta_f H[H_2O(g)]) = 2\ mol \times -241.8\ kJ/mol = -483.6\ kJ = \Delta H_2$

The heat needed to bring the products from 298 K to 500°C (or 773 K) is:
$\Delta H_3 = q = (2\ mol)(18.0\ g/mol)(1.864\ J/g{\cdot}K)(475\ K) = +31{,}870\ J$

The overall $\Delta_{rxn}H$ is the sum of these three parts. Converting all energy values to kJ:
$\Delta_{rxn}H = -41.131\ kJ - 483.6\ kJ + 31.87\ kJ = -491.9\ kJ$

CHAPTER 3. THE SECOND AND THIRD LAW OF THERMODYNAMICS

3.1. (a) Spontaneous, because ice's melting point is 0°C. (b) Not spontaneous, because ice's melting point is 0°C. (c) Spontaneous, because potassium compounds are generally soluble in water. (d) Not spontaneous; an unplugged refrigerator should warm up. (e) Spontaneous, because of the effect of gravity on the leaf. (f) Spontaneous, because both Li (s) and F_2 (g) are rather reactive elements. (g) Not spontaneous, because water does not break apart into hydrogen and oxygen without some input of energy.

3.3. $e = -\dfrac{w_{cycle}}{q_1} = -\dfrac{(-334-115+72+150)\,\text{J}}{850\,\text{J}} = -\dfrac{-227}{850} = 0.267 = 26.7\%$

3.5. $e = 1 - \dfrac{T_{low}}{T_{high}}$ $0.440 = 1 - \dfrac{T_{low}}{(150+273.15)\,\text{K}}$ $0.560 = \dfrac{T_{low}}{423\,\text{K}}$ $T_{low} = 237\,\text{K} = -36°\text{C}$

3.7. Superheated steam has the advantage of a higher temperature, so (hopefully) there will be a higher efficiency for a heat engine using superheated steam.

3.9. One definition of a perpetual motion machine is a machine that creates more energy than it uses. If it were to do so in an isolated system, it would violate the first law of thermodynamics. Perpetual motions of this sort are not known (and probably never will be).

3.11. All definitions of efficiency are applicable to real gases as well as ideal gases. Efficiency's definition is independent of the type of material involved in a process.

3.13. A better statement of the second law of thermodynamics includes the conditions under which the second law is strictly applicable: for an isolated system, a spontaneous change is always accompanied by an increase in entropy.

3.15. The units are standard, so let us simply substitute into the proper expression

$$\Delta S = 2.5 \times \int_{295\,\text{K}}^{1273\,\text{K}} \left(\frac{25.69 - 7.32 \times 10^{-4}T + 4.58 \times 10^{-6}T^2}{T} dT \right)$$

$$= 2.5 \times \int_{295\,\text{K}}^{1273\,\text{K}} \left(\frac{25.69}{T} - 7.32 \times 10^{-4} + 4.58 \times 10^{-6}T \right) dT$$

$$= 2.5 \times \left[25.69 \ln \frac{T_f}{T_i} \Big|_{295\,\text{K}}^{1273\,\text{K}} - 7.32 \times 10^{-4} T \Big|_{295\,\text{K}}^{1273\,\text{K}} + \frac{1}{2}(4.58 \times 10^{-6}T^2) \Big|_{295\,\text{K}}^{1273\,\text{K}} \right]$$

$$= 2.5 \times (37.56 - 0.7158 + 3.512) = 100.9\,\text{J/K}$$

3.17. The number of moles of air being breathed in depends, of course, on the temperature of the air. Let's assume a normal room temperature of 22°C = 295 K. (You may assume a slightly different temperature, but the final answer probably won't be too far off.) Under those conditions, 1 liter of air at 1 atm pressure is

$$n = \frac{pV}{RT} = \frac{(1\,\text{atm})(1\,\text{L})}{(0.08205\,\text{L} \cdot \text{atm/mol K})(295\,\text{K})} = 0.0413\,\text{mol air.}$$ The entropy change of 0.0413 mol

of gas undergoing a pressure change from 760 torr to 758 torr is

$$\Delta S = -nR\ln\frac{p_f}{p_i} = -(0.0413\,\text{mol})(8.314\,\text{J/mol} \cdot \text{K})\ln\frac{758\,\text{torr}}{760\,\text{torr}} = +9.05 \times 10^{-4}\,\text{J/K.}$$

3.19. Since the sample is a real gas, the change in entropy is probably greater than it would be than if it were an ideal gas. Therefore, let us calculate the entropy change assuming it is an ideal gas and state that the true entropy change must be greater than this. We separate the total entropy change into two parts, a change-in-pressure part and a change-in-volume part:

$$\Delta S_1 = -(1\,\text{mol})(8.314\,\text{J/mol} \cdot \text{K})\ln\frac{1\,\text{atm}}{230\,\text{atm}} = -45.2\,\text{J/K} \text{ for the change in pressure.}$$

$$\Delta S_2 = (1\,\text{mol})(8.314\,\text{J/mol} \cdot \text{K})\ln\frac{195\,\text{cm}^3}{1\,\text{cm}^3} = +43.8\,\text{J/K} \text{ for the change in volume.}$$

The overall entropy change is $\Delta S = -45.2 + 43.8 = -1.4$ J/K. Therefore, for our real gas we can suggest that the entropy change is probably greater than -1.4 J/K.

3.21. $5/2\,R$ is the heat capacity of an ideal gas under conditions of constant pressure. However, the process in Example 3.3 is not a constant-pressure process. Therefore, while the value of the heat capacity given may be the correct experimental value, for an ideal gas it probably wouldn't be.

3.23. The chemical processes can be represented as:

Ar (4.00 L, 298 K, 1.50 atm) + He (2.50 L, 298 K, 1.50 atm) → Ar, He (6.50 L, 298 K, 1.50 atm)

Ar, He (6.50 L, 298 K, 1.50 atm) → Ar, He (20.0 L, 298 K, 0.488 atm)

The new pressure was determined by simply applying Boyle's law. Since entropy is a state function, the change in entropy for the overall process can be determined by calculating the entropy changes for each process (a mixing process and an expansion process), then adding the two values. To determine the ΔS_{mix}, the number of moles and the mole fractions of Ar and He are needed:

$$\text{For Ar}: n = \frac{pV}{RT} = \frac{(1.50\,\text{atm})(4.00\,\text{L})}{(0.08205\,\text{L} \cdot \text{atm/mol} \cdot \text{K})(298\,\text{K})} = 0.245\,\text{mol}$$

$$\text{For He}: n = \frac{pV}{RT} = \frac{(1.50\,\text{atm})(2.50\,\text{L})}{(0.08205\,\text{L} \cdot \text{atm/mol} \cdot \text{K})(298\,\text{K})} = 0.153\,\text{mol}$$

The total number of moles of gas is $0.245 + 0.153 = 0.398$ mol, so the mole fractions of Ar and

He are $x_{\text{Ar}} = \frac{0.245}{0.398} = 0.616$, so $x_{\text{He}} = 1 - 0.616 = 0.384$.

The entropy change of the mixing step is therefore:

$\Delta S_{mix} = -(8.314 \text{ J/mol} \cdot \text{K})\left[(0.245 \text{ mol})\ln(0.61) + (0.153 \text{ mol})(\ln(0.384)\right] = +2.20 \text{ J/K}$

The entropy change for the expansion step is:

$\Delta S_{exp} = nR \ln \dfrac{V_f}{V_i} = (0.398 \text{ mol})(8.314 \text{ J/mol} \cdot \text{K}) \ln \dfrac{20.0 \text{ L}}{6.50 \text{ L}} = +3.72 \text{ J/K}$

Therefore, the total entropy of the process is $\Delta S = +2.20 + 3.72 = +5.92$ J/K.

3.25. (a) According to the first law, the energy of the isolated system remains constant. Therefore, any energy lost by one part of the system will be gained by another part of the system. In this case, the hot copper will lose energy and the cooler water will gain energy. In terms of the second law, this spontaneous change will occur only if the total entropy of the system increases. (b) Heat lost = heat gained. Note that heat lost is *negative*, while heat gained is *positive*.

Heat lost =

$-m \cdot c \cdot \Delta T = -m \cdot c \cdot (T_f - T_i) = -(5.33 \text{ g})(0.385 \dfrac{\text{J}}{\text{g} \cdot \text{K}})(T_f - 372.85 \text{ K}) = -2.052 T_f + 765.107$

Heat gained =

$+m \cdot c \cdot \Delta T = m \cdot c \cdot (T_f - T_i) = (99.53 \text{ g})(4.18 \dfrac{\text{J}}{\text{g} \cdot \text{K}})(T_f - 295.75 \text{ K}) = 416.03 T_f - 123,042.470$

Equate the two quantities and solve for T_f.

$-2.052 T_f + 765.107 = 416.03 T_f - 123,042.470$

$123,807.577 = 418.082 T_f$ 　　　　　　 $T_f = 296.13 \text{ K} = 23.0°\text{C}.$

(c) $\Delta S = mc \ln \dfrac{T_f}{T_i} = (5.33 \text{ g})(0.385 \text{ J/g} \cdot \text{K}) \ln \dfrac{296.13 \text{ K}}{372.85 \text{ K}} = -0.473 \text{ J/K} = \text{entropy loss of Cu}$

(d) $\Delta S = mc \ln \dfrac{T_f}{T_i} = (99.53 \text{ g})(4.18 \text{ J/g} \cdot \text{K}) \ln \dfrac{296.13 \text{ K}}{295.75 \text{ K}} = 0.534 \text{ J/K} = \text{entropy gain of water}$

(e) The total entropy change for the system is $\Delta S = -0.473 + 0.534 = 0.061$ J/K.
(f) Because the overall entropy change of the isolated system is positive, so we would expect that the process – the equalization of temperatures – would be spontaneous.

3.27. Since (for an isolated system) the first law of thermodynamics prohibits the creation of new energy, the concept of "you can't win" may be used to convey – if inaccurately – that fact. Since the second law of thermodynamics requires an efficiency of less than 100%, you will always get less energy out of a process than the energy going into that process. Thus, "you can't even break even" may be a way to convey that idea.

3.29. If $S = k \ln \Omega$, where Ω is the number of possible combinations, the number of combinations can never be less than 1 for any real system. The logarithm of 1 is zero, so it may be possible that the entropy of a system is zero. However, numbers greater than 1 have positive logarithms, and if k is a positive constant (which it is), then the absolute amount of entropy S can never be negative. (This does not preclude that *changes* in S might have negative values.)

3.31. (a) the dirty kitchen (b) the blackboard with writing on it (c) 10 gram of ice (d) If perfectly crystalline, both have the same entropy (zero) (e) 10 grams of ethanol at 22°C.

3.33. If helium is indeed a liquid at absolute zero, it would not have zero entropy. That's because according to the third law of thermodynamics, only a perfect crystal (that is, a solid) can have zero entropy at absolute zero.

3.35. Unlike the listings of $\Delta_f H$ and $\Delta_f G$, S for elements are not zero because the condition isn't what's required for S to equal zero: a perfect crystal at 0 K. Much of the tabulated thermodynamic data is for something close to room temperature, like 25°C (298 K).

3.37. The balanced chemical reaction is $2\ Al\ (s)\ +\ Fe_2O_3\ (s)\ \rightarrow\ Al_2O_3\ (s)\ +\ 2\ Fe\ (s)$
$\Delta S = [50.92 + 2(27.3)] - [2(28.30) + 87.4] = -38.5$ J/K

3.39. For the formation of H_2O (l), $\Delta S = 69.96 - 130.68 - \frac{1}{2}\,(205.14) = -163.29$ J/K. For the formation of H_2O (g), $\Delta S = 188.83 - 130.68 - \frac{1}{2}\,(205.14) = -44.42$ J/K. The difference is that the formation of the gas has a ΔS that is higher by 118.87 J/K. The reason for this difference is the different phase of the product H_2O.

3.41. $\Delta S = nC \ln \dfrac{T_f}{T_i} = (800\ \text{lb})(2200\ \text{g/lb})(0.45\ \text{J/g} \cdot \text{K}) \ln \dfrac{923\ \text{K}}{293\ \text{K}} = 9.09 \times 10^5$ J/K

3.43. At 37°C = 310 K, the number of moles of gas is
$n = \dfrac{pV}{RT} = \dfrac{(1\ \text{atm})(1\ \text{L})}{(0.08205\ \text{L} \cdot \text{atm/mol} \cdot \text{K})(310\text{K})} = 0.0393$ moles. The change in entropy is then
$\Delta S = -(0.0393\ \text{mol})(8.314\ \text{J/mol} \cdot \text{K}) \ln \dfrac{590\ \text{mmHg}}{760\ \text{mmHg}} = 0.0827$ J/K

CHAPTER 4. FREE ENERGY AND CHEMICAL POTENTIAL

4.1. Processes occur with change in energy as well as changes in entropy. Therefore, spontaneity conditions usually have to be determined with respect to both. However, if entropy changes are to be used as the sole, strict spontaneity condition, an isolated system is required, which prohibits changes in energy. Therefore both ΔU and ΔH must be zero.

4.3. The total change in U or H is required to be negative for these spontaneity conditions, not any one component (as given by a partial derivative).

4.5. Starting with the expression $dU + pdV - TdS \leq 0$, use the definition $dA = dU - TdS - SdT$ that we get by derivating eq. 4.5, solve for dU and substitute: $dU = dA + TdS + SdT$; therefore, $dA + TdS + SdT + pdV - TdS <= 0$, which simplifies to $dA + SdT + pdV \leq 0$. Under conditions of constant T and V, the second and third terms are zero, so this spontaneity condition simplifies to
$(dA)_{T,V} \leq 0$.

4.7. For internal energy, $dU = 0$ under conditions of constant volume and entropy.
For enthalpy, $dH = 0$ under conditions of constant pressure and entropy.
For Helmholtz energy, $dA = 0$ under conditions of constant volume and temperature.

4.9. The reaction can do up to 237.13 kJ of work for every mole of H_2O formed.

4.11. $\Delta U = \Delta H = 0$ (for an isothermal process). To determine the pressure at the bottom of the ocean, if the water pressure increases by 1 atm for every 10.55 meters of depth and the depth is 10,430 meters, then the pressure increase is $10{,}430\,\mathrm{m} \times \dfrac{1\,\mathrm{atm}}{10.55\,\mathrm{m}} = 989\,\mathrm{atm}$. If this is the pressure increase, then the pressure at that depth must be $989 + 1 = 990$ atm. Therefore:

$$w = -nRT \ln \frac{p_i}{p_f} = -(1\,\mathrm{mol})(8.314\,\mathrm{J/mol \cdot K})(273\,\mathrm{K}) \ln \frac{990\,\mathrm{atm}}{1\,\mathrm{atm}} = -15{,}700\,\mathrm{J}.$$

Since this is a reversible process, this equals ΔA as well. Finally, for ΔS:

$$\Delta S = nR \ln \frac{p_i}{p_f} = (1\,\mathrm{mol})(8.314\,\mathrm{J/mol \cdot K}) \ln \frac{990\,\mathrm{atm}}{1\,\mathrm{atm}} = +57.3\,\mathrm{J/K}.$$

4.13. For $NaHCO_3$ (s) $\rightarrow$ Na^+ (aq) $+$ HCO_3^- (aq), $\Delta G = [-261.88 + (-586.85)] - (-851.0) = +2.3$ kJ.
For Na_2CO_3 (s) $\rightarrow$ $2\,Na^+$ (aq) $+$ CO_3^{2-} (aq), $\Delta G = [2(-261.88) + (-386.0)] - (-1048.01) = +138.3$ kJ.

4.15. ΔG for the reaction is zero, because under the conditions given the phase change between liquid and solid water is an equilibrium. Data given in the appendix are given for 25°C, not 0°C. Solid H_2O is not thermodynamically stable at 25°C, so a lot of data is not given for that phase under that condition. However, since we know that 0°C is the normal melting point of H_2O under standard pressure, we expect that the value of ΔG would be exactly 0.

4.17. ΔA is zero for the complete Carnot cycle, since A is a state function and the cycle returns to its original conditions.

4.19. Analogous to equation 4.26:

$$A \equiv U - TS \qquad \therefore U = A + TS$$

One of the relationships we can determine about A is $\left(\dfrac{\partial A}{\partial T}\right)_V = -S$. We can substitute this for

the entropy variable in the second term: $U = A - T\left(\dfrac{\partial A}{\partial T}\right)_V$. The differential form of the equation

is

$$dU = dA - \left(\dfrac{\partial A}{\partial T}\right)_V dT \text{ for our final answer. We don't use the change of } A \text{ with respect to}$$

volume.

4.21. To show that $\left(\dfrac{\partial U}{\partial V}\right)_S = -p$ has consistent units on both sides of the equation, let us look at

the units distribution through the equation. U has units of energy (J), V has units of L, and p has

units of atm. Thus, we get $\dfrac{J}{L} = atm$, which does not appear to make much sense. But recall that

we can convert J to L·atm, so we can substitute L·atm for J in the numerator:

$\dfrac{L \cdot atm}{L} = \dfrac{atm}{1} = atm$, so the units are consistent.

4.23. (a) $\left(\dfrac{\partial}{\partial x}\left(\dfrac{\partial F}{\partial y}\right)\right) = \left(\dfrac{\partial}{\partial x}\left(\dfrac{\partial (x+y)}{\partial y}\right)\right) = \left(\dfrac{\partial}{\partial x}(1)\right) = 0$

$\left(\dfrac{\partial}{\partial y}\left(\dfrac{\partial F}{\partial x}\right)\right) = \left(\dfrac{\partial}{\partial y}\left(\dfrac{\partial (x+y)}{\partial x}\right)\right) = \left(\dfrac{\partial}{\partial y}(1)\right) = 0$. Since the two derivatives are equal, they are exact

differentials.

(b) $\left(\dfrac{\partial}{\partial x}\left(\dfrac{\partial F}{\partial y}\right)\right) = \left(\dfrac{\partial}{\partial x}\left(\dfrac{\partial (x^2+y^2)}{\partial y}\right)\right) = \left(\dfrac{\partial}{\partial x}(2y)\right) = 0$

$\left(\dfrac{\partial}{\partial y}\left(\dfrac{\partial F}{\partial x}\right)\right) = \left(\dfrac{\partial}{\partial y}\left(\dfrac{\partial (x^2+y^2)}{\partial x}\right)\right) = \left(\dfrac{\partial}{\partial y}(2x)\right) = 0$. Since the two derivatives are equal, they are

exact differentials.

(c) $\left(\dfrac{\partial}{\partial x}\left(\dfrac{\partial F}{\partial y}\right)\right) = \left(\dfrac{\partial}{\partial x}\left(\dfrac{\partial (x^n y^n)}{\partial y}\right)\right) = \left(\dfrac{\partial}{\partial x}(nx^n y^{n-1})\right) = n^2 x^{n-1} y^{n-1}$

$$\left(\frac{\partial}{\partial y}\left(\frac{\partial F}{\partial x}\right)\right) = \left(\frac{\partial}{\partial y}\left(\frac{\partial(x^n y^n)}{\partial x}\right)\right) = \left(\frac{\partial}{\partial y}\left(nx^{n-1}y^n\right)\right) = n^2 x^{n-1}y^{n-1}.$$ Since the two derivatives are equal, they are exact differentials.

(d) $$\left(\frac{\partial}{\partial x}\left(\frac{\partial F}{\partial y}\right)\right) = \left(\frac{\partial}{\partial x}\left(\frac{\partial(x^m y^n)}{\partial y}\right)\right) = \left(\frac{\partial}{\partial x}\left(nx^m y^{n-1}\right)\right) = mnx^{m-1}y^{n-1}$$

$$\left(\frac{\partial}{\partial y}\left(\frac{\partial F}{\partial x}\right)\right) = \left(\frac{\partial}{\partial y}\left(\frac{\partial(x^m y^n)}{\partial x}\right)\right) = \left(\frac{\partial}{\partial y}\left(mx^{m-1}y^n\right)\right) = mnx^{m-1}y^{n-1}.$$ Since the two derivatives are equal, they are exact differentials.

(e) $$\left(\frac{\partial}{\partial y}\left(\frac{\partial F}{\partial x}\right)\right) = \left(\frac{\partial}{\partial y}\left(\frac{\partial(y\sin(xy))}{\partial x}\right)\right) = \left(\frac{\partial}{\partial y}\left(y^2\cos(xy)\right)\right) = 2y\cos(xy) - y^2 x\sin(xy).$$

$$\left(\frac{\partial}{\partial x}\left(\frac{\partial F}{\partial y}\right)\right) = \left(\frac{\partial}{\partial x}\left(\frac{\partial(y\sin(xy))}{\partial y}\right)\right) = \left(\frac{\partial}{\partial x}(\sin(xy) + xy\cos(xy))\right) = y\cos(xy) + y\cos(xy) - xy^2\sin(xy)$$

Since the two derivatives are different, they are not exact differentials.

4.25. Starting with $dH = TdS + Vdp$, divide both sides by dp and hold T constant:

$$\left(\frac{\partial H}{\partial p}\right)_T = T\frac{\partial S}{\partial p} + V = V(1 + \frac{T}{V}\frac{\partial S}{\partial p})$$ All we need to do is show that $\alpha T = -\frac{1}{V}\frac{dS}{dp}$. Using the definition of α:

$$\alpha = \frac{1}{V}\left(\frac{\partial V}{\partial T}\right)_p = \frac{1}{V}\left(-\frac{\partial S}{\partial p}\right)_T$$, where we have used a Maxwell relation as a substitution.

Multiplying by T: $\alpha T = \frac{T}{V}\left(-\frac{\partial S}{\partial p}\right)_T$, which is the desired relationship. Therefore,

$$\left(\frac{\partial H}{\partial p}\right)_T = V(1 - \alpha T).$$

4.27. Simply put, by multiplying through by the denominator in the partial derivative, we get that $d(\Delta U) = -(\Delta p)dV$, which is a form of the definition for work. Recall that changes in U manifest in only two forms: as heat and/or as work. This expression is consistent with the work part of ΔU.

4.29. To demonstrate the cyclic rule of partial derivatives, let us evaluate the derivatives listed in Figure 1.11 using the ideal gas law:

$$\left(\frac{\partial p}{\partial T}\right)_V = \frac{\partial}{\partial T}\left(\frac{nRT}{V}\right) = \frac{nR}{V} \qquad \left(\frac{\partial V}{\partial T}\right)_p = \frac{\partial}{\partial T}\left(\frac{nRT}{p}\right) = \frac{nR}{p}$$

$$\left(\frac{\partial V}{\partial p}\right)_T = \frac{\partial}{\partial p}\left(\frac{nRT}{p}\right) = -\frac{nRT}{p^2}$$ and substitute:

$$\left(\frac{\partial p}{\partial T}\right)_V = -\frac{(\partial V / \partial T)_p}{(\partial V / \partial p)_T} \qquad \frac{nR}{V} = -\frac{(nR / p)}{-(nRT / p^2)} \qquad \text{which reduces to} \quad \frac{nR}{V} = +\frac{p}{T} \quad \text{or } pV = nRT,$$

verifying the cyclic rule.

4.31. Substituting for the definitions of α and κ:

$$\frac{\alpha}{\kappa}\left(\frac{\partial V}{\partial S}\right)_T = \frac{(1/V)\left(\frac{\partial V}{\partial T}\right)_p}{-(1/V)\left(\frac{\partial V}{\partial p}\right)_T}\left(\frac{\partial V}{\partial S}\right)_T . \quad \text{Using a Maxwell relationship, substitute for the final partial}$$

derivative, and rewrite the denominator as its reciprocal in the numerator (note that the $1/V$s cancel):

$$-\left(\frac{\partial V}{\partial T}\right)_p\left(\frac{\partial p}{\partial V}\right)_T\left(\frac{\partial T}{\partial p}\right)_V . \quad \text{Inspection of the three partial derivatives shows that the three variables}$$

are organized in such a way as to give the cyclic rule of partial differentiation, so that the product of the three derivatives is -1: $-(-1) = 1$.

4.33. According to a Maxwell relationship, $\left(\frac{\partial p}{\partial S}\right)_T = -\left(\frac{\partial T}{\partial V}\right)_p$. For an ideal gas,

$$-\left(\frac{\partial T}{\partial V}\right)_p = -\frac{\partial}{\partial V}\left(\frac{pV}{nR}\right)_p = -\frac{p}{nR} . \quad \text{For a van der Waals gas:}$$

$$T = \frac{\left(p + \frac{an^2}{V^2}\right)(V - nb)}{nR} \quad \text{Therefore :} \quad -\left(\frac{\partial T}{\partial V}\right)_p = -\frac{1}{nR}\left[\left(\frac{-2an^2}{V^3}\right)(V - nb) + \left(p + \frac{an^2}{V^2}\right)\right]. \quad \text{Here,}$$

we have had to take the derivative of the two V-containing terms using the chain rule.

4.35. The derivation of a Gibbs-Helmholtz expression for A is exactly the same as that given in section 4.7 in the text, except that A is substituted for G and U is substituted for H. Rather than redo the entire derivation, here we provide the final answer: $\frac{\partial}{\partial T}\left(\frac{A}{T}\right)_p = -\frac{U}{T^2}$, or

$$\frac{\partial}{\partial T}\left(\frac{\Delta A}{T}\right)_p = -\frac{\Delta U}{T^2} .$$

4.37. $\Delta G = nRT \ln\frac{p_f}{p_i} = nRT \ln\frac{V_i}{V_f} = (0.988 \text{ mol})(8.314 \text{ J/mol} \cdot \text{K})(350 \text{ K}) \ln\frac{25.0 \text{ L}}{35.0 \text{ L}} = -967 \text{ J}$.

The substitution for volumes instead of pressures was made using Boyle's law.

4.39. Starting with a natural variable expression for A, we know that A varies with volume as $\left(\frac{\partial A}{\partial V}\right)_T = -p$. We rewrite this as $dA = -pdV$. Integrating both sides: $\Delta A = -\int pdV$.

Substituting for p from the ideal gas law: $\Delta A = -\int \frac{nRT}{V} dV$, which integrates to

$\Delta A = -nRT \ln \frac{V_f}{V_i}$, which is the desired expression.

4.41. The change in the chemical potential is zero because the molar Gibbs free energy does not change for a single-phase, single-component system if the amount of the component changed. The molar Gibbs free energy, or chemical potential, is an intensive property. The total Gibbs free energy changes (because it is an extensive property), but μ does not.

4.43. $dG = -SdT + Vdp + \mu_{N_2} dn_{N_2} + \mu_{O_2} dn_{O_2}$. The first two terms can also be broken down into the individual entropies of the two components and the change in the pressures of the two components (assuming that dT and V are the same for the two gases).

4.45. 1.00 bar equals 0.987 atm. Therefore, going from 1.00 atm to 0.987 atm:

$\mu_{final} - \mu_{initial} = RT \ln \frac{p_f}{p_i} = (8.314 \text{ J/mol} \cdot \text{K})(273.15 \text{ K}) \ln \frac{0.987}{1.00} = -29.7 \text{ J}$. This may seem like a lot, but most chemical processes occur with energy changes in the thousands of joules, so the difference is relatively trivial.

4.47. (a) 1 mol of H_2O (g). (b) 10.0 g of Fe at 35°C. (c) the compressed air.

4.49. Oxygen should deviate from ideality more than helium at any pressure, although as the pressure decreases the behavior of both gases should approach that of an ideal gas.

CHAPTER 5. INTRODUCTION TO CHEMICAL EQUILIBRIUM

5.1. A battery that has a voltage is not at equilibrium because there still exists a driving force for a reaction to occur. On the other hand, a completely dead battery is at equilibrium because there is no driving force for a chemical reaction to occur.

5.3. (a) Rb^+ and OH^- and H_2 would be the prevalent equilibrium species. (b) NaCl (xtal) would be the prevalent equilibrium species. (c) H^+ (aq) and Cl^- (aq) would be the prevalent equilibrium species. (d) C (graphite) should be the prevalent equilibrium species, but from experience we know that C (diamond) is chemically very stable once formed.

5.5. The minimum value of ξ is 0 (as it is for any reaction). The maximum value can be found by determining which reagent is the limiting reagent:

$$100 \text{ g Zn} \times \frac{1 \text{ mol Zn}}{65.4 \text{ g}} = 1.53 \text{ mol Zn} \qquad 0.1500 \text{ L} \times \frac{2.25 \text{ mol HCl}}{\text{L}} = 0.3375 \text{ mol HCl}$$

The reaction requires a 1:2 molar ratio of Zn to HCl, but we have a 1:0.221 mole ratio, so HCl will be the limiting reagent. Therefore, according to the definition of ξ:

$$\xi = \frac{0 - 0.3375}{-2} = 0.1688 \text{ will be the maximum value of } \xi.$$

5.7. The heme + CO reaction lies farther towards products. In fact, one can show that the equilibrium constant for the reaction heme$\cdot O_2$ + CO $\longleftrightarrow$ heme$\cdot$CO + O_2 is $2.3 \times 10^{23}/9.2 \times 10^{18} = 2.5 \times 10^4$, indicating a strong preference by heme for CO – explaining its potential toxicity.

5.9. False. p° is the standard pressure for gaseous substances in a system, defined as either 1 atm or 1 bar.

5.11. (a) The equilibrium constant for this expression is $K = \dfrac{1}{p_{SO_2} \cdot p_{O_2}^{1/2}}$. SO_3 doesn't appear in the expression because it's in a condensed phase. (b) From the data in the appendix: $\Delta G^\circ = -368 - [-300.13 + \frac{1}{2}(0)] = -68$ kJ. (c) $\Delta G^\circ = -RT \ln K$. $-68,000 \text{ J} = -(8.314 \text{ J/mol} \cdot \text{K})(298.15 \text{ K}) \ln K$

$$\ln K = \frac{-68,000 \text{ J/mol}}{(-8.314 \text{ J/mol} \cdot \text{K})(298.15 \text{ K})} \qquad \ln K = 27.432 \qquad K = 8.2 \times 10^{11}. \text{ (d) The reaction}$$

would move toward the direction of more products.

5.13. For example, consider the following gas-phase reaction:

$$2 H_2 \text{ (g)} + O_2 \text{ (g)} \longleftrightarrow 2 H_2O \text{ (g)}$$

Its equilibrium constant expression is $K = \dfrac{p_{H_2O}^2}{p_{H_2}^2 \cdot p_{O_2}}$. Now, divide all coefficients by 2. We get

$$H_2 \text{ (g)} + \tfrac{1}{2} O_2 \text{ (g)} \longleftrightarrow H_2O \text{ (g)}$$

For this reaction, the equilibrium constant expression is $K = \dfrac{p_{H_2O}}{p_{H_2} \cdot p_{O_2}^{1/2}}$. The exponents on all of the partial pressures are half of what they were in the first expression, meaning that all of the terms in this expression are the square root of the terms in the original expression. Therefore, $K = (K)^{1/2}$ when the reaction itself is halved.

5.15. The reaction will reverse when ΔG equals 0. Let p be the pressure at which $\Delta G = 0$:

$0 = -32,800 + RT \ln \dfrac{p^2}{p \cdot p^3}$. Simplify and solve for p: $32,800 = (8.314)(298.15) \ln p^{-2}$

$\ln p^{-2} = \dfrac{32,800}{(8.314)(298.15)} = 13.232$ $\ln p = -6.6161$ $p = 1.34 \times 10^{-3}$ atm or bar for the reaction to reverse.

5.17. Addition of an inert gas to the equilibrium should not affect the position of the equilibrium because the gas does not participate in the reaction and the partial pressures of the gases involved in the reaction do not change.

5.19. $\Delta_{rxn}G^\circ = [-41.9 - 73.94] - [2(51.30 - 228.61] = 10.2$ kJ.

$10,200$ J/mol $= -(8.314$ J/mol $\cdot$ K)$(298.15$ K$) \ln K$ $\ln K = -\dfrac{10,200 \text{ J/mol}}{(8.314 \text{ J/mol} \cdot \text{K})(298.15 \text{ K})}$

$\ln K = -4.1148\ldots$ $K = 1.63 \times 10^{-2}$.

5.21. (a) $K = \left(\dfrac{\gamma_{Pb^{2+}} m_{Pb^{2+}}}{m^\circ} \right) \left(\dfrac{\gamma_{Cl^-} m_{Cl^-}}{m^\circ} \right)^2$ (b) $K = \dfrac{\left(\dfrac{\gamma_{H^+} m_{H^+}}{m^\circ} \right) \left(\dfrac{\gamma_{NO_2^-} m_{NO_2^-}}{m^\circ} \right)}{\left(\dfrac{\gamma_{HNO_2} m_{HNO_2}}{m^\circ} \right)}$

(c) $K = \dfrac{\left(\dfrac{p_{CO_2}}{p^\circ} \right)}{\left(\dfrac{\gamma_{H_2C_2O_4} m_{H_2C_2O_4}}{m^\circ} \right)}$.

5.23. Using the ΔG° from the previous problem: $\Delta G^\circ = -RT \ln \dfrac{a_{dia}}{a_{gra}}$, and substituting from equation 5.14:

2900 J/mol $= -RT \left[\dfrac{\overline{V}_{dia}}{RT}(p-1) - \dfrac{\overline{V}_{gra}}{RT}(p-1) \right]$. The RT terms cancel, leaving us with

2900 J/mol $= -\left[\overline{V}_{dia}(p-1) - \overline{V}_{gra}(p-1) \right]$. Now we need to substitute the molar volumes of diamond and graphite. Using 1 mol C = 12.01 g:

$$12.01 \, \text{g} \times \frac{1 \, \text{cm}^3}{2.25 \, \text{g}} \times \frac{1 \, \text{L}}{1000 \, \text{cm}^3} = 5.34 \times 10^{-3} \, \text{L for the molar volume of graphite.}$$

$$12.01 \, \text{g} \times \frac{1 \, \text{cm}^3}{3.51 \, \text{g}} \times \frac{1 \, \text{L}}{1000 \, \text{cm}^3} = 3.42 \times 10^{-3} \, \text{L for the molar volume of diamond. Substituting:}$$

$$2900 \, \text{J/mol} = -\left[(3.42 \times 10^{-3} \, \text{L/mol})(p-1) - (5.34 \times 10^{-3} \, \text{L/mol})(p-1)\right] \times \frac{101.32 \, \text{J}}{1 \, \text{L} \cdot \text{atm}} \text{ and solve for } p:$$

$p \approx 14{,}900$ atm.

5.25. (a) $\Delta G^\circ = -(8.314 \, \text{J/mol} \cdot \text{K})(298.15 \, \text{K}) \ln(1.2 \times 10^{-2})$ $\Delta G^\circ = 10.96$ kJ.
(b) Assuming that the salt is completely soluble, $[\text{Na}^+] = 0.010 \, m$. To determine the concentrations of HSO_4^-, H^+, and SO_4^{2-}, we first assume that HSO_4^- starts at a concentration of $0.010 \, m$, and that some amount $-x-$ dissociates, leaving $0.010 - x$. The amounts of SO_4^{2-} and H^+ are therefore $+x$. We therefore set up the following expression:

$K = 1.2 \times 10^{-2} = \dfrac{x \cdot x}{0.010 - x}$ and solve for x using the quadratic formula: $x = -0.0185$ or 0.00649.

We reject the negative root: $x = 0.00649$. Therefore, the final concentrations are $[\text{H}^+] = [\text{SO}_4^{2-}] = 0.00649 \, m$ and $[\text{HSO}_4^-] = 0.010 - 0.00649 = 0.00351 \, m$.

5.27. Use equation 5.20 and let $K_2 = 2K_1$:

$$\ln \frac{2K_1}{K_1} = \frac{-100{,}000 \, \text{J/mol}}{8.314 \, \text{J/mol} \cdot \text{K}} \left(\frac{1}{298 \, \text{K}} - \frac{1}{T_2} \right) \qquad \ln 2 = \frac{-100{,}000 \, \text{J/mol}}{8.314 \, \text{J/mol} \cdot \text{K}} \left(\frac{1}{298 \, \text{K}} - \frac{1}{T_2} \right) \quad \text{Solve for } T_2:$$

$$\frac{(\ln 2)(8.314)}{-100{,}000} = \frac{1}{298} - \frac{1}{T_2} \qquad T_2 = 293 \, \text{K.}$$

To determine a temperature for the equilibrium constant is 10 times the original value, let $K_2 = 10K_1$ and perform the same calculations:

$$\ln \frac{10K_1}{K_1} = \frac{-100{,}000 \, \text{J/mol}}{8.314 \, \text{J/mol} \cdot \text{K}} \left(\frac{1}{298 \, \text{K}} - \frac{1}{T_2} \right) \qquad \ln 10 = \frac{-100{,}000 \, \text{J/mol}}{8.314 \, \text{J/mol} \cdot \text{K}} \left(\frac{1}{298 \, \text{K}} - \frac{1}{T_2} \right) \quad \text{Solve for } T_2:$$

$$\frac{(\ln 10)(8.314)}{-100{,}000} = \frac{1}{298} - \frac{1}{T_2} \qquad T_2 = 282 \, \text{K.}$$

For a reaction whose $\Delta H = -20$ kJ, repeat both sets of calculations using this value of ΔH. Letting $K_2 = 2K_1$:

$$\ln \frac{2K_1}{K_1} = \frac{-20{,}000 \, \text{J/mol}}{8.314 \, \text{J/mol} \cdot \text{K}} \left(\frac{1}{298 \, \text{K}} - \frac{1}{T_2} \right) \qquad \ln 2 = \frac{-20{,}000 \, \text{J/mol}}{8.314 \, \text{J/mol} \cdot \text{K}} \left(\frac{1}{298 \, \text{K}} - \frac{1}{T_2} \right) \quad \text{Solve for } T_2:$$

$$\frac{(\ln 2)(8.314)}{-20{,}000} = \frac{1}{298} - \frac{1}{T_2} \qquad T_2 = 274 \, \text{K.}$$

To determine a temperature for the equilibrium constant is 10 times the original value, let $K_2 = 10K_1$ and perform the same calculations:

$$\ln \frac{10K_1}{K_1} = \frac{-20{,}000 \, \text{J/mol}}{8.314 \, \text{J/mol} \cdot \text{K}} \left(\frac{1}{298 \, \text{K}} - \frac{1}{T_2} \right) \qquad \ln 10 = \frac{-20{,}000 \, \text{J/mol}}{8.314 \, \text{J/mol} \cdot \text{K}} \left(\frac{1}{298 \, \text{K}} - \frac{1}{T_2} \right) \quad \text{Solve for } T_2:$$

$$\frac{(\ln 10)(8.314)}{-20,000} = \frac{1}{298} - \frac{1}{T_2} \qquad T_2 = 232 \text{ K.}$$

5.29. The easiest way to show that equations 5.18 and 5.19 are equivalent is to take the differential of $1/T$ and substitute:

$d(1/T) = -(1/T^2)dT$. Substitute this into equation 5.19:

$$\frac{d \ln K}{-(1/T^2)dT} = -\frac{\Delta H}{R} \text{ and bring the } -(1/T^2) \text{ term to the other side: } \frac{d \ln K}{dT} = -\left(\frac{1}{T^2}\right)\left(-\frac{\Delta H}{R}\right)$$

or $\dfrac{d \ln K}{dT} = \dfrac{\Delta H}{RT^2}$, which is equation 5.18.

5.31. If 1.0 mol of glycine is made into 1.00 L of solution, let us assume that the original pH is 7.00 and calculate the amount of protonated glycine is in the solution. According to equation 5.21:

$K_1 = \dfrac{[\text{gly}][\text{H}^+]}{[\text{glyH}^+]} = 10^{-2.34}$. If $[\text{gly}] = 1.00$ M and $[\text{H}^+] = 1.00 \times 10^{-7}$ M, we can calculate the

concentration of $[\text{glyH}^+]$: $10^{-2.34} = \dfrac{(1.00)(1.00 \times 10^{-7})}{[\text{glyH}^+]}$ $\qquad [\text{glyH}^+] = 2.19 \times 10^{-5}$ M .

Similarly, according to equation 5.21:

$K_2 = \dfrac{[\text{gly}^-][\text{H}^+]}{[\text{gly}]} = 10^{-9.60}$. Again, assuming $[\text{H}^+] = 1.00 \times 10^{-7}$ M, we can calculate the resulting

$[\text{gly}^-]$: $10^{-9.60} = \dfrac{[\text{gly}^-](1.00 \times 10^{-7})}{(1.00)}$ $\qquad [\text{gly}^-] = 2.51 \times 10^{-3}$ M .

CHAPTER 6. EQUILIBRIA FOR SINGLE COMPONENT SYSTEMS

6.1. (a) 1 component. (b) 2 components. (c) 4 components. (d) 2 components. (e) 2 components.

6.3. $FeCl_2$ and $FeCl_3$ are the only chemically-stable, single-component materials that can be made from iron and chlorine.

6.5. The water is boiling because the vapor pressure of the water equals the ambient pressure inside the syringe. By drawing back the plunger of the syringe enough, we can reduce the pressure on the water sufficiently so that the vapor pressure and ambient pressure are equal, which is the physical requirement for boiling.

6.7. Any substance has only one normal boiling point. The normal boiling point is the boiling point when the ambient pressure is 1.00 atm, and all substances only have one temperature where their vapor pressure is 1.00 atm.

6.9. (a) ΔH should be positive because energy has to go into a solid in order to sublime it. (b) ΔH should be negative because energy has to be removed from a gas to condense it.

6.11. The heat of fusion given up by the freezing water can be transferred (at least in part) to the citrus fruit, keeping them warmer and (hopefully) keeping the fruit itself from freezing.

6.13. $\Delta S = \dfrac{q_{rev}}{T} = \dfrac{+30,700 \text{ J}}{(80.1 + 273.15) \text{ K}} = +86.9 \text{ J/K}$, fairly close to what would be predicted by Trouton's rule.

6.15. Using the definition of entropy for this isothermal change: $\Delta S = \dfrac{q_{rev}}{T} = \dfrac{\Delta_{fus}H}{T}$ and

substituting: $124.7 \text{ J/mol} \cdot \text{K} = \dfrac{510,400 \text{ J/mol}}{T}$ $T = 4093 \text{ K or } 3820°C$.

6.17. The derivation of equation 6.12 from 6.11 assumes that the molar enthalpy change, $\Delta \overline{H}$, and the molar volume change, $\Delta \overline{V}$, do not vary with temperature.

6.19. First, we will need to calculate the molar volumes of rhombic and monoclinic sulfur:

$256.48 \text{ g} \times \dfrac{1 \text{ cm}^3}{2.07 \text{ g}} = 123.9 \text{ cm}^3 \times \dfrac{1 \text{ L}}{1000 \text{ cm}^3} = 0.1239 \text{ L}$ for the molar volume for rhombic sulfur

$256.48 \text{ g} \times \dfrac{1 \text{ cm}^3}{1.96 \text{ g}} = 130.9 \text{ cm}^3 \times \dfrac{1 \text{ L}}{1000 \text{ cm}^3} = 0.1309 \text{ L}$ for the molar volume for monoclinic sulfur

In going from monoclinic to rhombic sulfur, the change in molar volume is $0.1239 - 0.1309 \text{ L} = -0.0070 \text{ L}$. Using the Clapeyron equation: $\dfrac{\Delta p}{\Delta T} \approx \dfrac{\Delta \overline{S}}{\Delta \overline{V}} \rightarrow \dfrac{\Delta p}{-4.5 \text{ K}} = \dfrac{1.00 \text{ J/mol} \cdot \text{K}}{-0.0070 \text{ L}} \times \dfrac{1 \text{ L} \cdot \text{atm}}{101.32 \text{ J}}$

$\Delta \pi = \sim 6.3$ atm. Therefore, increasing the pressure by about 6.3 atm, to 7.3 atm, should be enough to make rhombic sulfur the stable form at 100°C.

6.21. (a) Yes, because a gas phase is involved. (b) Yes, because a gas phase is involved. (c) No, because no gas phase is involved. (d) No, because no gas phase is involved. (e) No, because the process isn't an equilibrium phase change. (f) No, because no gas phase is involved. (g) No, because no gas phase is involved. (h) Yes, because a gas phase is involved.

6.23. Using the fact that $\Delta \overline{V}$ was –0.0070 L for the phase change in exercise 6.19, and the fact that the monoclinic-to-rhombic phase change must have a ΔH of –0.368 kJ/mol:

$$\Delta p = \left(\frac{\Delta \overline{H}}{\Delta \overline{V}} \right) \ln \frac{T_f}{T_i} = \left(\frac{-368 \text{ J/mol}}{-0.0070 \text{ L/mol}} \right) \ln \frac{(100 + 273.15) \text{ K}}{(95.5 + 273.15) \text{ K}} \times \frac{1 \text{ L} \cdot \text{atm}}{101.32 \text{ J}} \qquad \Delta p = 6.3 \text{ atm, so that}$$

increasing the pressure by 6.3 atm to 7.3 atm should make the rhombic phase more stable. This is the same answer we got in exercise 6.19.

6.25. No, the behavior of chemical hot packs cannot be described using the Clapeyron equation or the Clausius-Clapeyron equation because supersaturated solutions are not equilibrium systems, nor does the process involve a phase change (it involves a solubility change).

6.27. If we use $\frac{dp}{dT} = \frac{\Delta H \cdot p}{RT^2}$ as the form of the Clausius-Clapeyron equation, we can substitute directly the values given in the problem:

$$\frac{dp}{dT} = \frac{(71,400 \text{ J/mol}) \cdot (7.9 \times 10^{-5} \text{ bar})}{(8.314 \text{ J/mol} \cdot \text{K})(22.0 + 273.15 \text{ K})^2} = 7.8 \times 10^{-6} \text{ bar/K}.$$

6.29. Since, according to equation 6.16, the vapor pressure is related to the exponential of the temperature, if the temperature is increased linearly, then the vapor pressure will increase exponentially. Thus, at high temperatures nearing the boiling point, even small changes in temperature can lead to large changes in vapor pressure.

6.31. (a) In order for the term γdA to have units of energy, and dA has units of m^2, γ must have units of J/m^2. (b) The derivative of A, in terms of r, is $dA = 8\pi r \, dr$ and the derivative of V in terms of r is $dV = 4\pi r^2 \, dr$. Substituting into the right side of the equation:

$$\frac{2dV}{r} = \frac{2(4\pi r^2 dr)}{r} = 8\pi r dr = dA.$$ (c) By simply rearranging the expression, we can get

$dV = \frac{dA \cdot r}{2}$. (d) According to the expression $dA = \frac{2dV}{r}$, droplets with smaller radii will contribute to a larger dA, which will in turn contribute to a larger value of dG. Since this expression relates the change in area to the change in volume – which relates directly to how fast the droplet is evaporating, we use this expression for dA, not the expression in part b. (e) According to our analysis in part d, smaller droplets should evaporate faster than large droplets. (f) See part e.

6.33. For equation 6.18, $\left(\dfrac{\partial \mu}{\partial T}\right)_{p,n} = -\overline{S}$, the units of μ are J/mol and the unit on T is K, so the overall unit on the left side is J/mol·K, which are the units for molar entropy. For equation 6.19, $\left(\dfrac{\partial \mu}{\partial p}\right)_{T,n} = \overline{V}$, the units of μ are J/mol and the unit on p is atm or bar, while the unit on the molar volume is L/mol. However, if we remember that the unit J can be written in terms of L·atm, then the units of μ can be written as L·atm/mol, which when divided by the unit atm equal units of L/mol, which are the units of molar volume.

6.35. Figure 6.6 shows 12 phase boundaries.

6.37. If there were such a thing as a single-axis phase diagram, equation 6.17 would probably be rewritten as 'degrees of freedom $= 2 - P$', because there would be one less variable needed to determine the exact state of the system.

6.39. The critical point represents the point in the phase diagram beyond which only a single fluid phase, a supercritical fluid, exists. Since $P = 1$, according to the Gibbs phase rule, degrees of freedom $= 3 - 1 = 2$, requiring two variables to be specified.

6.41. The answer will depend on the Figure and the line chosen. All of them are $\dfrac{\partial T}{\partial p}$ derivatives, but the phases involved vary with the line.

6.43. As sulfur goes from rhombic to monoclinic, the entropy should be increasing, just like it would if it were melting or vaporizing as the temperature is increased.

CHAPTER 7. EQUILIBRIA IN MULTIPLE COMPONENT SYSTEMS

7.1. There would be three degrees of freedom. They could include, for example, the mole fractions of alcohol, liquid water, and olive. The mole fraction of the fourth component can be determined by subtraction.

7.3. $0 = C - P + 2$ $0 = 3 - P + 2$ and solve for P: $P = 5$. So you would need 5 separate phases in equilibrium with each other to have zero degrees of freedom.

7.5. In this reaction, there are only three independent components (the amount of the fourth can be determined from the equilibrium constant) and 4 phases. Therefore, there are $3 - 4 + 2 = 1$ degree of freedom.

7.7. For water: $n = \dfrac{pV}{RT} = \dfrac{(23.76 \text{ torr})(5.00 \text{ L})}{(0.08205 \text{ L}\cdot\text{atm/mol}\cdot\text{K})(298.15 \text{ K})} \times \dfrac{1 \text{ atm}}{760 \text{ torr}} = 6.39 \times 10^{-3}$ mol of

water is needed to ensure that there is a liquid and gas phase. This equals 0.115 grams.

For methanol: $n = \dfrac{pV}{RT} = \dfrac{(125.0 \text{ torr})(5.00 \text{ L})}{(0.08205 \text{ L}\cdot\text{atm/mol}\cdot\text{K})(298.15 \text{ K})} \times \dfrac{1 \text{ atm}}{760 \text{ torr}} = 3.36 \times 10^{-2}$ mol of

methanol is needed to ensure that there is a liquid and gas phase. This equals 1.08 grams.

7.9. Since, according to equation 7.11, $a_i = \dfrac{p_i}{p_i^*}$, and since $p_i^* = 760$ torr, we can calculate a_i:

$$a_i = \dfrac{748.2 \text{ torr}}{760 \text{ torr}} = 0.984$$

7.11. We start with the expressions similar to those in equation 7.18 in the text, but in terms of y_2:

$y_2 = \dfrac{p_2}{p_{tot}} = \dfrac{p_2}{p_2 + p_1} = \dfrac{x_2 p_2^*}{x_2 p_2^* + x_1 p_1^*}$. Next, we recognize that $x_1 = 1 - x_2$, so we substitute in the

denominator: $y_2 = \dfrac{x_2 p_2^*}{x_2 p_2^* + (1 - x_2) p_1^*}$. This rearranges algebraically (see exercise 7.10) into

$y_2 = \dfrac{x_2 p_2^*}{p_1^* + (p_2^* - p_1^*) x_2}$, which is the equation of interest.

7.13. $p(\text{ethanol}) = (0.0006)(115.5 \text{ torr}) = 0.0693 \text{ torr} = 0.07 \text{ torr}$ (1 sig fig)

7.15. Equation 7.19 is $y_1 = \dfrac{x_1 p_1^*}{p_2^* + (p_1^* - p_2^*) x_1}$. First, let us multiply through the x_1 in the

denominator: $y_1 = \dfrac{x_1 p_1^*}{p_2^* + x_1 p_1^* - x_1 p_2^*}$. Now, multiply the denominator over to the other side of the equation: $y_1(p_2^* + x_1 p_1^* - x_1 p_2^*) = x_1 p_1^*$, or $y_1 p_2^* + y_1 x_1 p_1^* - y_1 x_1 p_2^* = x_1 p_1^*$. Collect all terms having x_1 in it on one side of the equation: $y_1 p_2^* = x_1 p_1^* - y_1 x_1 p_1^* + y_1 x_1 p_2^*$ and factor the x_1 out of all three terms on the right side: $y_1 p_2^* = x_1(p_1^* - y_1 p_1^* + y_1 p_2^*)$. Now, divide the parenthetical terms over to the other side of the equation (which we will flip):

$x_1 = \dfrac{y_1 p_2^*}{p_1^* - y_1 p_1^* + y_1 p_2^*}$. Now factor out the y_1s from the last two terms in the denominator:

$x_1 = \dfrac{y_1 p_2^*}{p_1^* + (p_2^* - p_1^*)y_1}$, which is our ultimate expression.

7.17. Consider the equation $p_{tot} = \dfrac{p_2^* p_1^*}{p_1^* + (p_2^* - p_1^*)y_1}$. If y_1 goes to zero, then the entire second term in the denominator is zero, and the expression becomes $\dfrac{p_2^* p_1^*}{p_1^*}$, which reduces to p_2^*, as it should when the system contains only component #2. If y_1 goes to 1, then the expression for p_{tot} becomes $\dfrac{p_2^* p_1^*}{p_1^* + p_2^* - p_1^*}$, which becomes $\dfrac{p_2^* p_1^*}{p_2^*} = p_1^*$, as it should when the system contains only component #1.

7.19. $\Delta_{mix}G = (2 \text{ mol})(8.314 \text{ J/mol·K})(293.15 \text{ K})[(0.5)(\ln 0.5) + (0.5)(\ln 0.5)] = -3380 \text{ J}$
$\Delta_{mix}S = -(2 \text{ mol})(8.314 \text{ J/mol·K})[(0.5)(\ln 0.5) + (0.5)(\ln 0.5)] = +11.5 \text{ J/K}$.

7.21. If an initial composition of $x_1 = 0.1$ were used, the tie lines that you would draw would eventually lead you to the minimum-boiling azeotrope, near the middle of the phase diagram. Thus, the azeotrope will be your ultimate product.

7.23. An azeotrope can be distinguished from a pure component by determining its temperature profile as it freezes. If it is a pure component, it will have a distinct freezing point for the entire sample. However, if it is two (or more) components, then each component should freeze at distinctly different temperatures.

7.25. The mixture of water and ethylene glycol has both a higher boiling point and lower freezing point than pure water alone.

7.27. Hydrogen chloride is a diatomic gas, whereas hydrochloric acid is HCl that has been dissolved in water. Since HCl is a strong acid and virtually 100% ionized in solution, it is doubtful that a solution of HCl will act ideally. HCl is very polar and unlikely to act ideally.

7.29. A mole fraction of 4.17×10^{-5} implies that there are 4.17×10^{-5} moles of CCl_2F_2 in 0.9999583 moles of water. Assuming no volume change when such a small amount of solute is added to water, the volume of 0.9999583 moles of water is

$$0.9999583 \, \text{mol} \times \frac{18.01 \, \text{g}}{1 \, \text{mol}} \times \frac{1.00 \, \text{cm}^3}{1.00 \, \text{g}} = 18.0 \, \text{cm}^3, \text{ or } 0.0180 \, \text{L}.$$ Therefore the molarity of this

solution is $\dfrac{4.17 \times 10^{-5} \, \text{mol}}{0.0180 \, \text{L}} = 0.00232 \, \text{M}$. To determine the Henry's law constant, use 1 atm =

101,325 Pa as the pressure:

$$101{,}325 \, \text{Pa} = K_i \times (4.17 \times 10^{-5}) \qquad K_i = 2.43 \times 10^{9} \, \text{Pa}.$$

7.31. (a) The mole fraction of nitrogen in water is about 90% of the mole fraction of air in water. Since nitrogen is 80% of air, it should not be surprising that the majority of the mole fraction of air in water is composed of nitrogen. (b) If air is 20% oxygen and the mole fraction of air in water is 1.388×10^{-5}, let us assume that oxygen would be 20% of that mole fraction (assuming that the composition of the air dissolved in water is the same as the composition of gaseous air). Therefore, the mole fraction of oxygen in water would be about $(0.20)(1.388 \times 10^{-5})$ $= 2.78 \times 10^{-6}$. (c) If oxygen is 20% of 1 atm = 101,325 Pa, then the partial pressure of oxygen is $(0.20)(101{,}325 \, \text{Pa}) = 20{,}265 \, \text{Pa}$. Therefore: $20{,}265 \, \text{Pa} = K_i \times (2.78 \times 10^{-6}) \qquad K_i = 7.29 \times 10^{9} \, \text{Pa}$. This is the same answer we got in exercise 7.30, and it should be – all we did was multiply both numerical values by 0.20, so the calculated value of K_i should be the same. However, Table 7.1 shows that the actual value of K_i is somewhat less than this $(4.34 \times 10^{9} \, \text{Pa})$, indicating that nitrogen and oxygen do not dissolve in water to an extent proportional to their composition of air.

7.33. If 87.0 grams of phenol can be dissolved in 100 mL of water, to calculate the molarity we need to know the moles of phenol and the total volume of the solution. The number of moles of

phenol are $87.0 \, \text{g} \times \dfrac{1 \, \text{mol}}{94 \, \text{g}} = 0.925 \, \text{mol phenol}$. The volume of the phenol is given by

$87.0 \, \text{g} \times \dfrac{1 \, \text{mL}}{1.06 \, \text{g}} = 82.1 \, \text{mL}$. If we assume that the volumes are additive, then the total volume of

the solution is 100 + 82.1 mL = 182.1 mL = 0.1821 L. Therefore, the molarity of the solution is $\dfrac{0.925 \, \text{mol}}{0.1821 \, \text{L}} = 5.08 \, \text{M}$.

7.35. (a) The calculated mole fraction of naphthalene in toluene is 0.311, so there are 0.311 mol of naphthalene in 0.689 mol of toluene. The molecular weight of toluene $(C_6H_5CH_3)$ is 92.0 g/mol, and using the density of toluene as given, we can calculate the volume of toluene used:

$0.689 \, \text{mol} \times \dfrac{92.0 \, \text{g}}{1 \, \text{mol}} \times \dfrac{1 \, \text{mL}}{0.866 \, \text{g}} = 73.2 \, \text{mL}$ of volume. We do the same for naphthalene $(C_{10}H_8)$:

$0.311 \, \text{mol} \times \dfrac{128.0 \, \text{g}}{1 \, \text{mol}} \times \dfrac{1 \, \text{mL}}{1.025 \, \text{g}} = 38.8 \, \text{mL}$ of volume. Thus, the total volume is 73.2 + 38.8 =

112.0 mL = 0.1120 L. Determining the molarity: $M = \dfrac{0.311 \, \text{mol}}{0.112 \, \text{L}} = 2.78 \, \text{M}$. (b) Because the

ideal solubility is calculated using only properties of the solute, the calculated mole fraction solubility of naphthalene in n-decane is the same as in toluene: $x_{\text{solute}} = 0.311$. However, the concentrations expressed in other units will be different. To get the solubility in grams per 100 mL of solvent:

$$\frac{0.311\,\text{mol}}{0.689\,\text{mol}} \times \frac{128.0\,\text{g}\,C_{10}H_8}{\text{mol}} \times \frac{1\,\text{mol}}{142.0\,\text{g decane}} \times \frac{0.730\,\text{g}}{1\,\text{mL}} = \frac{39.8\,\text{g naphthalene}}{134\,\text{mL decane}} \times \frac{100/134}{100/134} = \frac{29.7\,\text{g}}{100\,\text{mL}}$$

In terms of molarity, again we need to determine the total volume of the two components:

$$0.689\,\text{mol} \times \frac{142.0\,\text{g}}{1\,\text{mol}} \times \frac{1\,\text{mL}}{0.730\,\text{g}} = 134.0\,\text{mL of volume}.$$ We do the same for naphthalene ($C_{10}H_8$):

$$0.311\,\text{mol} \times \frac{128.0\,\text{g}}{1\,\text{mol}} \times \frac{1\,\text{mL}}{1.025\,\text{g}} = 38.8\,\text{mL of volume}.$$ Thus, the total volume is $134.0 + 38.8 =$

172.8 mL = 0.1728 L. Thus, the molarity is: $M = \dfrac{0.311\,\text{mol}}{0.1728\,\text{L}} = 1.80\,\text{M}$.

7.37. Of the four solutions listed, the $C_{20}H_{42}$ in cyclohexane is probably the closest to ideal. The sodium chloride/water and sucrose/water solutions deviate because of polar interactions between solute and solvent, and the water/carbon tetrachloride combines a polar solute with a nonpolar solvent. Therefore, the calculated properties of the $C_{20}H_{42}$/cyclohexane solution will probably be closest to actual properties.

7.39. $\ln(8.0 \times 10^{-3}) = -\dfrac{14,900\,\text{J/mol}}{8.314\,\text{J/mol}\cdot\text{K}}\left(\dfrac{1}{298.15\,\text{K}} - \dfrac{1}{T}\right) \quad -4.828314... = -1792\left(\dfrac{1}{298.15} - \dfrac{1}{T}\right)$

$0.002694 = (0.003354 - 1/T) \quad 1/T = 0.006600.... \quad T = 1515\,\text{K} = 1242°\text{C}.$

7.41. We can tell by looking at a phase diagram of the NaCl-H$_2$O system and see if the temperatures and relative concentrations involved point to a eutectic or to the colligative property. But as mentioned in the chapter, the percent of NaCl in the salt/water eutectic is 23% salt (that is, about 1 part NaCl to 3 parts H$_2$O). Personal experience suggests that salting roads doesn't use that much salt with respect to water, so the melting phenomenon is actually due to the colligative property of freezing-point depression.

7.43. The drawing is left to the student.

7.45. $\ln x_{\text{solute}} = -\dfrac{2,600\,\text{J/mol}}{8.314\,\text{J/mol}\cdot\text{K}}\left(\dfrac{1}{273.15\,\text{K}} - \dfrac{1}{(97.8 + 273.15)\,\text{K}}\right) \quad \ln x = -0.30184...$

$x = 0.739.$

7.47. Molarity includes the concept of partial molar volume because it is defined using the number of liters *of solution*. Thus, even if partial molar volumes may change during a range of compositions, only the total volume is used to define molarity.

7.49. First, let us calculate K_f and K_b:

$$K_f = \frac{M_{\text{solv}} R T_{\text{MP}}{}^2}{1000\Delta_{\text{fus}}H} = \frac{(18.02\,\text{g/mol})(8.314\,\text{J/mol}\cdot\text{K})(273.15\,\text{K})^2}{(1000\,\text{g/kg})(6009\,\text{J/mol})} = 1.86\,\text{K/molal}$$

29

$$K_b = \frac{M_{solv}RT_{BP}^2}{1000\Delta_{vap}H} = \frac{(18.02 \text{ g/mol})(8.314 \text{ J/mol} \cdot \text{K})(373.15 \text{ K})^2}{(1000 \text{ g/kg})(40,660 \text{ J/mol})} = 0.513 \text{ K/molal}$$

For the freezing point depression: $\Delta T = (2 \text{ particles})(1.86 \text{ K/molal})(1.08 \text{ molal}) = 4.02 \text{ K}$. Therefore, the freezing point goes down by $4.02°$: FP = -4.02°C.

For the boiling point elevation: $\Delta T = (2 \text{ particles})(0.513 \text{ K/molal})(1.08 \text{ molal}) = 1.11 \text{ K}$. Therefore, the boiling point goes up by $1.11°$: BP = 101.11°C.

For the osmotic pressure, we need the mole fraction of the solute. If there is 1.08 moles of "NaCl" per liter of solution (yielding 2.16 mol of particles per liter), and a liter of solution contains approximately 1000 mL of water (assuming negligible volume change from the solute), that's 55.5 moles of H_2O. Therefore, the mole fraction of solute is $\frac{2.16}{57.65} = 0.0375$. The molar volume of the solution is approximately the same as the molar volume of water, or 0.01802 L. Using the van't Hoff equation, equation 7.56:

$$\Pi \overline{V} = xRT \qquad \Pi(0.01802 \text{ L}) = (0.0375)(0.08314 \text{ L} \cdot \text{bar/mol} \cdot \text{K})(298.15\text{K})$$

Solve for Π: 51.5 bar.

7.51. If $x = 0.739$, we can use the van't Hoff equation:
$$\Pi \overline{V} = xRT \qquad \Pi(0.0152 \text{ L}) = (0.739)(0.08314 \text{ L} \cdot \text{bar/mol} \cdot \text{K})(273.15\text{K})$$

Solve for Π: $\Pi = 1210$ bar.

7.53. $K_f = \dfrac{M_{solv}RT_{MP}^2}{1000\Delta_{fus}H} = \dfrac{(159.8 \text{ g/mol})(8.314 \text{ J/mol} \cdot \text{K})(273.15-7.2 \text{ K})^2}{(1000 \text{ g/kg})(10,570 \text{ J/mol})} = 8.89 \text{ K/molal}$

$K_b = \dfrac{M_{solv}RT_{BP}^2}{1000\Delta_{vap}H} = \dfrac{(159.8 \text{ g/mol})(8.314 \text{ J/mol} \cdot \text{K})(273.15+58.78 \text{ K})^2}{(1000 \text{ g/kg})(29,560 \text{ J/mol})} = 4.95 \text{ K/molal}$

7.55. $m = \dfrac{n}{\# \text{ kg}} \approx \dfrac{x_i}{\text{L water}} = \dfrac{\Pi}{RT}$ $\qquad m = \dfrac{30 \text{ Pa}}{(0.08314 \text{ L} \cdot \text{bar/mol} \cdot \text{K})(310 \text{ K})} \times \dfrac{1 \text{ bar}}{100,000 \text{ Pa}}$

$\underline{m} = 1.16 \times 10^{-5}$ molal. For 1 kg of solvent (water):

$$1.16 \times 10^{-5} \text{ molal} \times 1 \text{ kg} \times \frac{185,000 \text{ g}}{\text{mol}} = 2.15 \text{ g polymer}.$$

CHAPTER 8. ELECTROCHEMISTRY AND IONIC SOLUTIONS

8.1. $F = \dfrac{q_1 q_2}{4\pi\varepsilon_0 r^2}$ $0.0225\,\text{N} = \dfrac{q_1(1.00\,\text{C})}{4\pi(8.854\times10^{-12}\,\text{C}^2/\text{J}\cdot\text{m})(100.0\,\text{m})^2}$ $q_1 = 2.50\times10^{-8}\,\text{C}$

8.3. (a) $F = \dfrac{q_1 q_2}{4\pi\varepsilon_0 r^2}$ $1.55\times10^{-6}\,\text{N} = \dfrac{2q^2}{4\pi(78)(8.854\times10^{-12}\,\text{C}^2/\text{J}\cdot\text{m})(0.06075\,\text{m})^2}$ Solve for q:

$q = 4.98\times10^{-9}$ C, so one particle has that charge and the other has twice that, or 9.96×10^{-9} C. (b) The electric fields for the two particles are equal to the force divided by their charges.

Therefore, the electric field on the first particle is $\dfrac{1.55\times10^{-6}\,\text{N}}{4.98\times10^{-9}\,\text{C}} = 311\,\text{J/C}\cdot\text{m}$, while the electric

field on the other particle is $\dfrac{1.55\times10^{-6}\,\text{N}}{9.96\times10^{-9}\,\text{C}} = 156\,\text{J/C}\cdot\text{m}$.

8.5. $F = \dfrac{(+1.602\times10^{-19}\,\text{C})(-1.602\times10^{-19}\,\text{C})}{4\pi(8.854\times10^{-12}\,\text{C}^2/\text{J}\cdot\text{m})(5.29\times10^{-11}\,\text{m})^2}$ $F = 8.24\times10^{-8}$ N. This may not be
much force, but it's huge compared to the sizes of the proton and electron!

8.7. "Electromotive force" is not a force in that it is not a mass times an acceleration, or even the equivalent. Rather, "electromotive force" is a difference between two electric potentials, which have units of J/C and not newtons.

8.9. (a) The two half reactions are:

$MnO_2 + 2\,H_2O \rightarrow MnO_4^- + 4\,H^+ + 3\,e^-$ $E = -1.679$ V
$4\,e^- + O_2 + 2\,H_2O \rightarrow 4\,OH^-$ $E = 0.401$ V

To balance the electrons, we multiply the first reaction by 4 and the second reaction by 3. Combining the H^+ and OH^- ions to H_2O and consolidating the water molecules on both sides, we get

$4\,MnO_2 + 3\,O_2 + 2\,H_2O \rightarrow 4\,MnO_4^- + 4\,H^+$ $E = -1.278$ V

To calculate ΔG°:

$\Delta G^\circ = -nFE = -(12\,\text{mol})(96{,}485\,\text{C/mol})(-1.278\,\text{V}) = +1.480\times10^6\,\text{J} = 1480\,\text{kJ}$

(b) The two half reactions are:

$Cu^+ \rightarrow Cu^{2+} + e^-$ $E = -0.153$ V
$e^- + Cu^+ \rightarrow Cu$ $E = 0.521$ V

Because both half reactions have one electron, they can be combined directly without multiplying through either reaction. The overall reaction is

$2\,Cu^+ \rightarrow Cu + Cu^{2+}$ $E = 0.368$ V

To calculate ΔG°:

$\Delta G^\circ = -nFE = -(1\,\text{mol})(96{,}485\,\text{C/mol})(0.368\,\text{V}) = -3.55\times10^4\,\text{J} = -35.5\,\text{kJ}$

(c) The two half reactions are:

$Br_2 + 2\,e^- \rightarrow 2\,Br^-$ $E = 1.087$ V
$2\,F^- \rightarrow F_2 + 2\,e^-$ $E = -2.866$ V

Because both half reactions have one electron, they can be combined directly without multiplying through either reaction. The overall reaction is

$$Br_2 + 2\,F^- \rightarrow F_2 + 2\,Br^- \quad E = -1.779\ V$$

To calculate $\Delta G°$:

$$\Delta G° = -nFE = -(2\ mol)(96{,}485\ C/mol)(-1.779\ V) = +3.433 \times 10^5\ J = 343.3\ kJ$$

(d) The two half reactions are:

$$H_2O_2 + 2\,H^+ + 2\,e^- \rightarrow 2\,H_2O \qquad E = 1.776\ V$$
$$2\,Cl^- \rightarrow Cl_2 + 2\,e^- \qquad E = -1.358\ V$$

Because both half reactions have two electrons, they can be combined directly without multiplying through either reaction. The overall reaction is

$$2\,H_2O_2 + 2\,H^+ + 2\,Cl^- \rightarrow 2\,H_2O + Cl_2 \qquad E = 0.418\ V$$

To calculate $\Delta G°$:

$$\Delta G° = -nFE = -(2\ mol)(96{,}485\ C/mol)(0.418\ V) = -8.07 \times 10^4\ J = -80.7\ kJ$$

8.11. The two half reactions are:

$$Fe^{2+} \rightarrow Fe^{3+} + e^- \qquad E = -0.771\ V$$
$$Fe^{2+} + 2\,e^- \rightarrow Fe \qquad E = -0.447\ V$$

To balance the electrons, we multiply the first reaction by 2. For the overall reaction, we get

$$3\,Fe^{2+} \rightarrow 2\,Fe^{3+} + Fe \qquad E = -1.218\ V$$

Since the overall voltage is negative, the reaction is not spontaneous. To calculate $\Delta G°$:

$$\Delta G° = -nFE = -(2\ mol)(96{,}495\ C/mol)(-1.218\ V) = 2.350 \times 10^5\ J = 235.0\ kJ$$

8.13. For the standard hydrogen electrode, the spontaneous process would be

$$2\,Li + 2\,H^+ \rightarrow 2\,Li^+ + H_2$$

The voltage of this process is 3.04 V. For the standard calomel electrode, the process is

$$2\,Li + Hg_2Cl_2 \rightarrow 2\,Li^+ + 2\,Hg + 2\,Cl^-$$

and the voltage is 3.31 V. Therefore, when the calomel half reaction is used as the reduction reaction, the voltage shifts up by 0.2682 V. However, consider the other half reaction with silver. With the hydrogen electrode, the spontaneous process is

$$2\,Ag + 2\,H^+ \rightarrow 2\,Ag^+ + H_2$$

and the voltage is 0.7996 V. With the calomel electrode, the spontaneous reaction is

$$2\,Ag + 2\,Hg + 2\,Cl^- \rightarrow 2\,Ag^+ + Hg_2Cl_2$$

and the voltage is 0.5314 V, so the voltage shifts down by 0.2682 V. Therefore, the direction of the shift depends on how the electrode reaction is used.

8.15. Because these half-reactions typically occur in an aqueous solvent, the interactions between the ionic species and the solvent molecules has an impact on the overall energy change (in terms of E and ΔG) of the process. Although all alkali metal ions have a +1 charge, the smaller, higher-charge-density lithium ion interacts more strongly with water molecules, increasing the energy of the process.

8.17. $E = E° - \dfrac{RT}{nF} \ln \dfrac{[Zn^{2+}]}{[Cu^{2+}]}$ $E°$ for this reaction is 1.1037 V, so we have

$$1.000 \text{ V} = 1.1037 \text{ V} - \frac{(8.314 \text{ J/K})(298.15 \text{ K})}{(2 \text{ mol})(96,485 \text{ C/mol})} \ln \frac{[\text{Zn}^{2+}]}{[\text{Cu}^{2+}]}$$ This rearranges to

$\ln \frac{[\text{Zn}^{2+}]}{[\text{Cu}^{2+}]} = 8.07279...$ $\quad \frac{[\text{Zn}^{2+}]}{[\text{Cu}^{2+}]} = 3.21 \times 10^3$. Unfortunately, we can't mathematically

determine specific concentrations without more information; we can only specify the ratio.

8.19. (a) $E° = 0$ V for any concentration cell, since both half reactions are the same but opposite.

(b) $Q = \frac{[\text{Fe}^{3+}]}{[\text{Fe}^{3+}]} = \frac{0.001}{0.08}$. (c) $E = 0 \text{ V} - \frac{(8.314 \text{ J/K})(298.15 \text{ K})}{(3 \text{ mol})(96,485 \text{ C/mol})} \ln \frac{0.001}{0.08} = 0.0375 \text{ V}$. (d) The

opinion is left to the student.

8.21. In order to use the equation $E \approx E° + \left(\frac{\Delta S°}{nF} \right) \Delta T$, we will need the entropy change of the

reaction. The entropy of H^+ (aq) is defined as zero, and we are assuming that $S(\text{D}^+, \text{aq})$ is zero also. Therefore, the entropy change of the reaction is (using data from the appendix) 144.96 − 130.68 = 14.28 J/K. The number of electrons transferred is 2 so we have:

$0 \approx -0.044 \text{ V} + \left(\frac{14.28 \text{ J/K}}{(2 \text{ mol})(96,485 \text{ C/mol})} \right) \Delta T$ Solving for ΔT: $\Delta T = 595$ K. Therefore, if we

raise the temperature from 298 K to $(298 + 595) = \sim 893$ K, the voltage of the reaction should be about 0.

8.23. Since heat capacity (at constant pressure) is defined as $\left(\frac{\partial H}{\partial T} \right)_p$, we can take the derivative

of equation 8.30 with respect to temperature: $\left(\frac{\partial (\Delta H)}{\partial T} \right)_p = -nF \left(\frac{\partial E°}{\partial T} + \left[\frac{\partial E°}{\partial T} + T \frac{\partial^2 E°}{\partial T^2} \right] \right)$,

which simplifies to $\Delta C_p = -nF \left(2 \frac{\partial E°}{\partial T} + T \frac{\partial^2 E°}{\partial T^2} \right)$.

8.25. $E = 0 \text{ V} - \frac{(8.314 \text{ J/K})(298.15 \text{ K})}{(2 \text{ mol})(96,485 \text{ C/mol})} \ln \frac{0.0077}{0.035} = 0.0194 \text{ V}$

8.27. Following Example 8.7: the reaction can be written in terms of two half reactions:

$$\text{AgCl (s)} + \text{e}^- \rightarrow \text{Ag (s)} + \text{Cl}^- \text{ (aq)} \qquad E = 0.22233 \text{ V}$$
$$\text{Ag (s)} \rightarrow \text{Ag}^+ + \text{e}^- \qquad E = -0.7996 \text{ V}$$

The overall voltage of the combination of the two reactions is −0.5773 V. Using equation 8.32:

$-0.5773 \text{ V} = \frac{(8.314 \text{ J/K})(298.15 \text{ K})}{(1 \text{ mol})(96,485 \text{ C/mol})} \ln K_{\text{sp}}$ Solve for K_{sp}: $K_{\text{sp}} = 1.74 \times 10^{-10}$.

8.29. First, we need to determine the overall reaction. Using Table 8.2, finding the MnO_4^-/Mn^{2+} half reaction, and combining it with the hydrogen electrode reaction, we find a 10-electron overall reaction:

$$2\,MnO_4^- + 6\,H^+ + 5\,H_2 \rightarrow 2\,Mn^{2+} + 8\,H_2O \qquad E^\circ = 1.507\ V$$

However, because some of the concentrations are not standard, equation 8.35 can't be used directly. It is probably best to use the complete Nernst equation:

$$1.200\ V = 1.507\ V - \frac{(8.314\ J/K)(298.15\ K)}{(10\ mol)(96{,}485\ C/mol)}\ln\frac{(0.288)^2}{(0.034)^2[H^+]^6(1)^5} \quad \text{Solving for } [H^+]:$$

$$7.88\times10^{51} = \frac{(0.288)^2}{(0.034)^2[H^+]^6(1)^5} \quad [H^+] = 4.6\times10^{-9} \quad \text{Therefore, the pH} = -\log(4.6\times10^{-9}) = 8.34.$$

8.31. The K_{sp} for Hg_2Cl_2 was determined in exercise 8.28 and is 1.29×10^{-18}. If x moles per liter of Hg_2Cl_2 dissociates, one gets x M Hg_2^{2+} and $2x$ M Cl^-. Therefore, we have:
$$1.29\times10^{-18} = (x)(2x)^2 \qquad 1.29\times10^{-18} = 4x^3 \qquad x = 6.86\times10^{-7} \quad \text{Since the equilibrium concentration}$$
of Cl^- is twice this, $[Cl^-] = 1.38\times10^{-6}$ M.

8.33. Using the definition of ionic strength in equation 8.47:

(a) $I = \frac{1}{2}\left((0.0055m)(+1)^2 + (0.0055m)(-1)^2\right) = 0.0055m$

(b) $I = \frac{1}{2}\left((0.075m)(+1)^2 + (0.075m)(-1)^2\right) = 0.075m$

(c) $I = \frac{1}{2}\left((0.0250m)(+2)^2 + (0.0500m)(-1)^2\right) = 0.0750m$

(d) $I = \frac{1}{2}\left((0.0250m)(+3)^2 + (0.0750m)(-1)^2\right) = 0.150m$

8.35. In the equation H_2 (g) + I_2 (s) $\rightarrow$ 2 H^+ (aq) + 2 I^- (aq), the overall enthalpy of reaction is -110.38 kJ. Using the concept of products-minus-reactants to determine ΔH, we need the heats of formation of the products (one of which is the object of this calculation) and the heats of formation of the reactants. The two reactants are elements, so their $\Delta_f H$s are zero. By convention, the $\Delta_f H$ of H^+ (aq) is also zero, so the only non-zero $\Delta_f H$ is that for I^-. Therefore, we have
$$-110.38\ kJ = (2\ mol)(\Delta_f H[I^-]) \qquad \Delta_f H\ [I^-] = -55.19\ kJ/mol.$$

8.37. The reaction is HF (g) $\rightarrow$ H^+ (aq) + F^- (aq). Using the thermodynamic values from the appendix, we have:
$\Delta H = (-332.63 + 0) - (-273.30) = -59.33$ kJ
$\Delta S = (-13.8 + 0) - (173.779) = -187.6$ J/K
$\Delta G = (-278.8 + 0) - (-274.6) = -4.2$ kJ
(b) Using $\Delta G^\circ = -RT\ln K$: -4200 J/mol = $-(8.314$ J/mol·K)(298.15 K) ln K
ln $K = 1.6943....$ $K = 5.44$. This is rather far off from the 3.5×10^{-4} value as measured. The difference is that the equilibrium constant refers to HF in the aqueous phase being the reactant,

rather than the gas phase. The predicted value for K should be much closer to the measured value if the hydration of HF step were included.

8.39. The complete expression for A is worked out in the text. The student need simply verify that the numbers and units do reduce to 1.171 molal$^{-1/2}$.

8.41. 0.9% NaCl implies 0.9 g NaCl in 99.1 g water. Assuming that the volume of the solution is 100 mL = 0.100 L = 0.100 kg: $\dfrac{(0.9\,\text{g})(1\,\text{mol}/58.5\,\text{g})}{0.100\,\text{kg}} = 0.154$ molal. Therefore, the ionic strength is $I = \dfrac{1}{2}\left((0.154\,\text{molal})(+1)^2 + (0.154\,\text{molal})(-1)^2\right) = 0.154$ molal.

8.43. (a) Identity of the counterion is necessary to determine the ionic strengths of the solutions.
(b) If the counterion were sulfate instead of nitrate, the ionic strengths would need to be recalculated. If sulfate were the counterion, the salt's formula is $Fe_2(SO_4)_3$ and the sulfate concentration is 3/2 of the iron ion concentration:
$I\,(Fe^{3+}\,\text{soln}) = \frac{1}{2}\cdot[(0.100)(+3)^2 + (0.150)(-2)^2] = 0.750$ molal
If sulfate were the counterion, the other salt's formula would be $CuSO_4$ and the sulfate concentration would equal the copper ion concentration:
$I\,(Cu^{2+}\,\text{soln}) = \frac{1}{2}\cdot[(0.050)(+2)^2 + (0.050)(-2)^2] = 0.200$ molal
Using equation 8.52 for each ion:
$$\ln\gamma_{Fe^{3+}} = -\frac{(1.171\,\text{molal}^{-1/2})(+3)^2(0.750\,\text{molal})^{1/2}}{1+(2.32\times10^9\,\text{m}^{-1}\text{molal}^{-1/2})(9.00\times10^{-10}\,\text{m})(0.750\,\text{molal})^{1/2}}$$
$\ln\gamma_{Fe^{3+}} = -3.250$ $\qquad \gamma\,(Fe^{3+}) = 0.0388$ $\qquad$ Therefore, the activity of Fe^{3+} is $(0.0388)(0.100\ m)$
$= 0.00388\ m$.
Similarly, for the Cu^{2+}:
$$\ln\gamma_{Cu^{2+}} = -\frac{(1.171\,\text{molal}^{-1/2})(+2)^2(0.200\,\text{molal})^{1/2}}{1+(2.32\times10^9\,\text{m}^{-1}\text{molal}^{-1/2})(6.00\times10^{-10}\,\text{m})(0.200\,\text{molal})^{1/2}}$$
$\ln\gamma_{Cu^{2+}} = -1.291$ $\qquad \gamma\,(Cu^{2+}) = 0.275$ $\qquad$ Therefore, the activity of Cu^{2+} is $(0.275)(0.050\ m) =$
$0.0138\ m$.
Substituting these activities into the Nernst equation:
$$E = 0.379\ \text{V} + \frac{(8.314\ \text{J/K})(298.15\text{K})}{(6\ \text{mol})(95{,}485\ \text{C/mol})}\ln\frac{(0.00388)^2}{(0.0138)^3} = 0.379 + 0.0075 = 0.386\ \text{V}$$
When you compare this to the voltage from the example (0.372 V), we see how the ionic strength of the solution, as influenced by the counterion, can have an influence on the voltage even though the counterions don't participate in the reaction.

8.45. Equation 8.61 is $I = e^2 \cdot |z|^2 \cdot \left(\dfrac{N_i}{V}\right) \cdot A \cdot \dfrac{E}{6\pi\eta r_i}$. Equation 8.5 shows that E has units of

N/C, e has units of C, z is unitless (it is simply the magnitude of the charge on the ion), the fraction (N_i/V) has units of $1/\text{m}^3$, A has units of m^2, and in SI units, viscosity has units of $\text{kg/m}\cdot\text{s}$. The numbers 6 and π have no units. The radius r has units of m. Combining all of these units:

$C^2 \cdot \left(\dfrac{1}{m^3} \right) \cdot m^2 \cdot \dfrac{N/C}{(kg/m \cdot s)(m)}$ This can be rearranged to get $\dfrac{C^2 \cdot m^2 \cdot N \cdot m \cdot s}{C \cdot m^3 \cdot m \cdot kg}$.   One of the C

units cancels, as do the three m units in the numerator and three of the four m units in the

denominator. This results in $\dfrac{C \cdot N \cdot s}{kg \cdot m}$. If we break down the newton unit into kg·m/s², we have

$\dfrac{C \cdot kg \cdot m \cdot s}{s^2 \cdot kg \cdot m}$. The kg and m units cancel, as do one of the second units in the numerator and

denominator. What's left is C/s, and one coulomb per second is an ampere, the unit of current.

8.47. For a galvanic cell, oxidation (the loss of electrons) occurs at the cathode and is considered
the negative electrode. Therefore, I_- is the current towards the cathode, and I_+ must be the
current towards the anode. In an electrolytic cell, oxidation (the loss of electrons) still occurs at
the cathode, so I_- is still the current towards the cathode and I_+ is the current towards the anode.
For any given cell reaction, however, the identities of the cathode and anode are switched for
galvanic and electrolytic cells.

CHAPTER 9. PRE-QUANTUM MECHANICS

9.1. The kinetic energy for a mass falling in the z direction is $\frac{1}{2}m\dot{z}^2$ and the gravitational potential energy is mgz. Therefore, the Lagrangian is $L = \frac{1}{2}m\dot{z}^2 - mgz$. Using the formula $\frac{d}{dt}\left(\frac{\partial L}{\partial \dot{z}}\right) = \frac{\partial L}{\partial z}$, we determine the following derivatives:

$$\left(\frac{\partial L}{\partial \dot{z}}\right) = \frac{\partial}{\partial \dot{z}}\left(\frac{1}{2}m\dot{z}^2 - mgz\right) = m\dot{z}$$

$$\frac{d}{dt}\left(\frac{\partial L}{\partial \dot{z}}\right) = \frac{d}{dt}(m\dot{z}) = m\ddot{z}$$

$$\frac{\partial L}{\partial z} = \frac{\partial}{\partial z}\left(\frac{1}{2}m\dot{z}^2 - mgz\right) = -mg$$

Combining the last two lines, we have for the Lagrangian equation of motion: $m\ddot{z} = -mg$.

9.3. In this case, q is z, $\dot{q}$ is $\dot{z}$, p is $-m\dot{z}$, and $\dot{p}$ is $-m\ddot{z}$ (these last two are because an object falling in the z direction is decreasing its z value). According to equations 9.14 and 9.15, $\left(\frac{\partial H}{\partial m\dot{z}}\right) = \dot{z}$ and $\left(\frac{\partial H}{\partial z}\right) = m\ddot{z}$. The first differential is easy to demonstrate:

$$\left(\frac{\partial H}{\partial m\dot{z}}\right) = \frac{1}{m}\left(\frac{\partial H}{\partial \dot{z}}\right) = \frac{1}{m}\frac{\partial}{\partial \dot{z}}(\frac{1}{2}m\dot{z}^2 + mgz) = \frac{1}{m}(m\dot{z} + 0) = \dot{z}$$, as required. The derivative of the

second term is zero because $\dot{z}$ does not show up as a variable in that term. The second differential is a little tricky. If we perform the differentiation, we get

$$\left(\frac{\partial H}{\partial z}\right) = \frac{\partial}{\partial z}(\frac{1}{2}m\dot{z}^2 + mgz) = (0 + mg) = mg$$, which does not appear to be the required

expression. However, remember what g is: it is the acceleration due to gravity experienced by the mass falling down. Therefore, g IS $\ddot{z}$. Therefore, we do in fact have $\left(\frac{\partial H}{\partial z}\right) = m\ddot{z} = -\dot{p}$, as required by Hamilton's laws of motion.

9.5. The drawing is left to the student.

9.7. If two spectra of two different compounds have some lines at exactly the same wavelength, then by Kirchhoff's and Bunsen's proposition, they must share one or more constituent element.

9.9. To determine the series limit, assume that $1/n_1^2 = 1/\infty^2 = 0$. For the Lyman series:

wavenumber $= 109,700 \text{ cm}^{-1}\left(\frac{1}{1^2} - 0\right) = 109,700 \text{ cm}^{-1}$ is the series limit. For the Brackett series:

wavenumber $= 109,700 \text{ cm}^{-1}\left(\frac{1}{4^2} - 0\right) = 6856 \text{ cm}^{-1}$ is the series limit.

9.11. First, we need to convert the wavelengths to wavenumbers:

$$656.2 \text{ nm} \times \frac{1 \text{ m}}{10^9 \text{ nm}} \times \frac{100 \text{ cm}}{1 \text{ m}} = 6.562 \times 10^{-5} \text{ cm} \qquad \therefore \frac{1}{\lambda} = \frac{1}{6.562 \times 10^{-5} \text{ cm}} = 15,240 \text{ cm}^{-1}$$

$$486.1 \text{ nm} \times \frac{1 \text{ m}}{10^9 \text{ nm}} \times \frac{100 \text{ cm}}{1 \text{ m}} = 4.861 \times 10^{-5} \text{ cm} \qquad \therefore \frac{1}{\lambda} = \frac{1}{4.861 \times 10^{-5} \text{ cm}} = 20,570 \text{ cm}^{-1}$$

$$434.0 \text{ nm} \times \frac{1 \text{ m}}{10^9 \text{ nm}} \times \frac{100 \text{ cm}}{1 \text{ m}} = 4.340 \times 10^{-5} \text{ cm} \qquad \therefore \frac{1}{\lambda} = \frac{1}{4.340 \times 10^{-5} \text{ cm}} = 23,040 \text{ cm}^{-1}$$

Using these wavenumber values, we can calculate R for $n_2 = 2$, assuming that the first three lines correspond to $n_2 = 3$, 4, and 5, respectively:

$$15,240 \text{ cm}^{-1} = R\left(\frac{1}{2^2} - \frac{1}{3^2}\right) \quad 15,240 \text{ cm}^{-1} = R(0.1388888...) \quad R = 109,730 \text{ cm}^{-1}$$

$$20,570 \text{ cm}^{-1} = R\left(\frac{1}{2^2} - \frac{1}{4^2}\right) \quad 20,570 \text{ cm}^{-1} = R(0.1875) \quad R = 109,710 \text{ cm}^{-1}$$

$$23,040 \text{ cm}^{-1} = R\left(\frac{1}{2^2} - \frac{1}{5^2}\right) \quad 23,040 \text{ cm}^{-1} = R(0.21) \quad R = 109,710 \text{ cm}^{-1}$$

The average of these three is $\dfrac{109,730 + 109,710 + 109,710}{3} = 109,720 \text{ cm}^{-1}$.

9.13. (a) A single electron has a mass of 9.109×10^{-31} kg. If we assume that a helium nucleus has the mass of two protons plus two neutrons (ignoring the mass defect that represents the energy stabilization of the nucleus), then the helium nucleus has a mass of $(2 \times 1.672 \times 10^{-27} + 2 \times 1.675 \times 10^{-27})$ kg $= 6.694 \times 10^{-27}$ kg. The ratio of these two masses will tell us how many beta particles it takes to equal the mass of 1 alpha particle: $\dfrac{6.694 \times 10^{-27}}{9.109 \times 10^{-31}} = 7349$ beta particles to mass the same as one alpha particle. (b) A beta particle would be moving faster than an alpha particle of the same kinetic energy, since it has a much smaller mass. (c) This is consistent with the fact that beta particles are known to be the more penetrating particle.

9.15. For a flux of 1.00 W/m^2: $1.00 \dfrac{W}{m^2} = (5.6705 \times 10^{-8} \dfrac{W}{m^2 K^4})(T^4)$ Solving for T : $T = 65$ K

For a flux of 10.00 W/m^2: $10.00 \dfrac{W}{m^2} = (5.6705 \times 10^{-8} \dfrac{W}{m^2 K^4})(T^4)$ Solving for T : $T = 115$ K

For a flux of 100.00 W/m^2: $100.00 \dfrac{W}{m^2} = (5.6705 \times 10^{-8} \dfrac{W}{m^2 K^4})(T^4)$ Solving for T : $T = 205$ K

9.17. (a) power per unit area $= (5.6705 \times 10^{-8} \text{ W/m}^2 K^4)(5800 \text{ K})^4 = 6.42 \times 10^7 \text{ W/m}^2$

(b) $6.42 \times 10^7 \text{ W/m}^2 \times 6.087 \times 10^{12} \text{ m}^2 = 3.91 \times 10^{20}$ W

(c) 3.91×10^{20} W $= 3.91 \times 10^{20}$ J/s $\times \dfrac{365\,\text{d}}{1\,\text{year}} \times \dfrac{24\,\text{hr}}{1\,\text{d}} \times \dfrac{3600\,\text{s}}{1\,\text{hr}} = 1.23 \times 10^{28}$ J per year.

9.19. (a) Using Wien's law: $\lambda_{max} \cdot 5800$ K $= 2898$ µm·K $\lambda_{max} = 0.4997$ µm $= 4997$Å.
(b) 5000 Å $= 0.5000$ µm, so using Wien's law: 0.5000 µm·$T = 2898$ µm·K $T = 5796$ K. (c)
The two are very close, suggesting (but not proving!) that the eye may have evolved to take
advantage of the brightest part of the sun's spectrum.

9.21. Planck's law is written as $dE = \dfrac{2\pi hc^2}{\lambda^5}\left(\dfrac{1}{e^{hc/\lambda kt}-1}\right)d\lambda$. In order to use the given integral,

we're going to have to redefine the variable for frequency, not wavelength. Since $c = \lambda\nu$, we

have $\lambda = \dfrac{c}{\nu}$ and $d\lambda = -\dfrac{c}{\nu^2}d\nu$. Substituting:

$dE = \dfrac{2\pi hc^2}{\lambda^5}\left(\dfrac{1}{e^{hc/\lambda kt}-1}\right)d\lambda = \dfrac{2\pi hc^2}{(c/\nu)^5}\left(\dfrac{1}{e^{hc/(c/\nu)kT}-1}\right)\left(-\dfrac{c}{\nu^2}d\nu\right)$ This expression simplifies into

$dE = -\dfrac{2\pi h\nu^3}{c^2}\left(\dfrac{1}{e^{h\nu/kT}-1}\right)d\nu$. If we now substitute $x \equiv h\nu/kT$ so that $dx = h/kT \cdot d\nu$, or

$d\nu = kT/h \cdot dx$ and $\nu = xkT/h$, we substitute again:

$dE = -\dfrac{2\pi h(xkT/h)^3}{c^2}\left(\dfrac{1}{e^x-1}\right)\dfrac{kT}{h}dx$. Collecting terms: $dE = -\dfrac{2\pi k^4 T^4}{h^3 c^2}\cdot\dfrac{x^3}{e^x-1}$. Now we can

integrate from $x = 0$ to $x = $ (you should verify that when $\nu = 0$, $x = 0$, and when $\nu = $, $x = $:

$E = \int dE = \int_0^{\infty}-\dfrac{2\pi k^4 T^4}{h^3 c^2}\cdot\dfrac{x^3}{e^x-1}\,dx = -\dfrac{2\pi k^4 T^4}{h^3 c^2}\int_0^{\infty}\dfrac{x^3}{e^x-1}\,dx = -\dfrac{2\pi k^4 T^4}{h^3 c^2}\cdot\dfrac{\pi^4}{15} = -\dfrac{2\pi^5 k^4 T^4}{15 h^3 c^2}$ which is

the Stefan-Boltzmann law. (The negative sign indicates that energy is being given off.)

9.23. Substituting the values of the various constants:
$\dfrac{2\pi^5 (1.381 \times 10^{-23}\,\text{J/K})^4}{15(2.9979 \times 10^8\,\text{m/s})^2 (6.626 \times 10^{-34}\,\text{Js})^3} = 5.676 \times 10^{-8}$ W/m^2 , which is almost exactly the value
of the Stefan-Boltzmann constant quoted in the text.

9.25. A work function of 2.16 eV is equal to $2.16\,\text{eV} \times \dfrac{1.602 \times 10^{-19}\,\text{J}}{1\,\text{eV}} = 3.46 \times 10^{-19}$ J.

Determine the energy equivalent of the wavelengths of light, then subtract 3.46×10^{-19} J from that
energy value. Any remaining energy is converted to kinetic energy of motion, $\frac{1}{2}mv^2$.

(a) $550\,\text{nm} \times \dfrac{1\,\text{m}}{10^9\,\text{nm}} = 5.50 \times 10^{-7}$ m $\nu = \dfrac{2.9979 \times 10^8\,\text{m/s}}{5.50 \times 10^{-7}\,\text{m}} = 5.45 \times 10^{14}\,\text{s}^{-1}$

$E = (6.626 \times 10^{-34}\,\text{Js})(5.45 \times 10^{14}\,\text{s}^{-1})$
$E = 3.61 \times 10^{-19}$ J.
If 3.46×10^{-19} J are used in overcoming the work function, there is 0.15×10^{-19} J left. Using the
formula for kinetic energy and the mass of the electron:

$$0.15 \times 10^{-19} \text{ J} = \frac{1}{2}(9.109 \times 10^{-31} \text{ kg})v^2 \qquad v = 181{,}000 \text{ m/s}.$$

(b)
$$450 \text{ nm} \times \frac{1 \text{ m}}{10^9 \text{ nm}} = 4.50 \times 10^{-7} \text{ m} \qquad v = \frac{2.9979 \times 10^8 \text{ m/s}}{4.50 \times 10^{-7} \text{ m}} = 6.66 \times 10^{14} \text{ s}^{-1}$$

$$E = (6.626 \times 10^{-34} \text{ Js})(6.66 \times 10^{14} \text{ s}^{-1})$$

$$E = 4.41 \times 10^{-19} \text{ J}.$$

If 3.46×10^{-19} J are used in overcoming the work function, there is 0.95×10^{-19} J left. Using the formula for kinetic energy and the mass of the electron:

$$0.95 \times 10^{-19} \text{ J} = \frac{1}{2}(9.109 \times 10^{-31} \text{ kg})v^2 \qquad v = 457{,}000 \text{ m/s}.$$

(c)
$$350 \text{ nm} \times \frac{1 \text{ m}}{10^9 \text{ nm}} = 3.50 \times 10^{-7} \text{ m} \qquad v = \frac{2.9979 \times 10^8 \text{ m/s}}{3.50 \times 10^{-7} \text{ m}} = 8.57 \times 10^{14} \text{ s}^{-1}$$

$$E = (6.626 \times 10^{-34} \text{ Js})(8.57 \times 10^{14} \text{ s}^{-1})$$

$$E = 5.68 \times 10^{-19} \text{ J}.$$

If 3.46×10^{-19} J are used in overcoming the work function, there is 2.22×10^{-19} J left. Using the formula for kinetic energy and the mass of the electron:

$$2.22 \times 10^{-19} \text{ J} = \frac{1}{2}(9.109 \times 10^{-31} \text{ kg})v^2 \qquad v = 698{,}000 \text{ m/s}.$$

9.27. By combining the equations $E = h\nu$ and $c = \lambda\nu$, we get an expression that can be used directly: $E = \dfrac{hc}{\lambda}$. For the wavelength of 10 m:

$$E = \frac{(6.626 \times 10^{-34} \text{ J} \cdot \text{s})(2.9979 \times 10^8 \text{ m/s})}{10 \text{ m}} = 1.99 \times 10^{-26} \text{ J per photon}, \times 6.022 \times 10^{23} / \text{mol} = 0.0120 \text{ J/mol photons}$$

For a wavelength of 10.0 cm = 0.100 m:

$$E = \frac{(6.626 \times 10^{-34} \text{ J} \cdot \text{s})(2.9979 \times 10^8 \text{ m/s})}{0.100 \text{ m}} = 1.99 \times 10^{-24} \text{ J per photon}, \times 6.022 \times 10^{23} / \text{mol} = 1.20 \text{ J/mol photons}$$

For a wavelength of 10 microns = 0.00001 m:

$$E = \frac{(6.626 \times 10^{-34} \text{ J} \cdot \text{s})(2.9979 \times 10^8 \text{ m/s})}{0.00001 \text{ m}} = 1.99 \times 10^{-20} \text{ J per photon}, \times 6.022 \times 10^{23} / \text{mol} = 12000 \text{ J/mol photons}$$

For a wavelength of 550 nm = 5.50×10^{-7} m:

$$E = \frac{(6.626 \times 10^{-34} \text{ J} \cdot \text{s})(2.9979 \times 10^8 \text{ m/s})}{5.50 \times 10^{-7} \text{ m}} = 3.61 \times 10^{-19} \text{ J per photon}, \times 6.022 \times 10^{23} / \text{mol} = 217{,}500 \text{ J/mol photons}$$

For a wavelength of 300 nm = 3.00×10^{-7} m:

$$E = \frac{(6.626 \times 10^{-34} \text{ J} \cdot \text{s})(2.9979 \times 10^8 \text{ m/s})}{3.00 \times 10^{-7} \text{ m}} = 6.62 \times 10^{-19} \text{ J per photon}, \times 6.022 \times 10^{23} / \text{mol} = 398{,}700 \text{ J/mol photons}$$

For a wavelength of 1 Å = 10^{-10} m:

$$E = \frac{(6.626 \times 10^{-34} \text{ J} \cdot \text{s})(2.9979 \times 10^8 \text{ m/s})}{10^{-10} \text{ m}} = 1.99 \times 10^{-15} \text{ J per photon}, \times 6.022 \times 10^{23} / \text{mol} = 1.20 \times 20^9 \text{ J/mol photons}$$

1.20×10^9

9.29. Using equation 9.34, for $n = 4$:

40

$$r = \frac{\varepsilon_0 n^2 h^2}{\pi m_e e^2} = \frac{(8.854 \times 10^{-12} \text{ C}^2/\text{Jm})(4^2)(6.626 \times 10^{-34} \text{ J} \cdot \text{s})^2}{\pi (9.109 \times 10^{-31} \text{ kg})(1.602 \times 10^{-19} \text{ C})^2} = 8.47 \times 10^{-10} \text{ m} = 8.47 \text{ A}$$

For $n = 5$:

$$r = \frac{\varepsilon_0 n^2 h^2}{\pi m_e e^2} = \frac{(8.854 \times 10^{-12} \text{ C}^2/\text{Jm})(5^2)(6.626 \times 10^{-34} \text{ J} \cdot \text{s})^2}{\pi (9.109 \times 10^{-31} \text{ kg})(1.602 \times 10^{-19} \text{ C})^2} = 1.32 \times 10^{-9} \text{ m} = 13.2 \text{ A}$$

For $n = 6$:

$$r = \frac{\varepsilon_0 n^2 h^2}{\pi m_e e^2} = \frac{(8.854 \times 10^{-12} \text{ C}^2/\text{Jm})(6^2)(6.626 \times 10^{-34} \text{ J} \cdot \text{s})^2}{\pi (9.109 \times 10^{-31} \text{ kg})(1.602 \times 10^{-19} \text{ C})^2} = 1.91 \times 10^{-9} \text{ m} = 19.1 \text{ A}$$

9.31. Angular momenta are simply integral units of $h/2\pi$: for $n = 4$,
$L = 4 \cdot (6.626 \times 10^{-34} \text{ J} \cdot \text{s})/2\pi = 4.22 \times 10^{-34} \text{ J} \cdot \text{s}$.
For $n = 5$: $L = 5 \cdot (6.626 \times 10^{-34} \text{ J} \cdot \text{s})/2\pi = 5.27 \times 10^{-34} \text{ J} \cdot \text{s}$.
For $n = 6$: $L = 6 \cdot (6.626 \times 10^{-34} \text{ J} \cdot \text{s})/2\pi = 6.33 \times 10^{-34} \text{ J} \cdot \text{s}$.

9.33. (a) $v = \dfrac{e^2}{2\varepsilon_0 nh}$ (b) For $n = 1$,

$$v = \frac{(1.602 \times 10^{-19} \text{ C})^2}{2(8.854 \times 10^{-12} \text{ C}^2/\text{Jm})(1)(6.626 \times 10^{-34} \text{ Js})} = 2.18 \times 10^6 \text{ m/s}$$ This is about 0.73% of the speed

of light. (c) $L = mvr = 9.109 \times 10^{-31} \text{ kg} \cdot 2.19 \times 10^6 \text{ m/s} \cdot 5.29 \times 10^{-11} \text{ m} = 1.06 \times 10^{-34} \text{ J} \cdot \text{s}$, which is equal to $h/2\pi$.

9.35. Starting with $n\lambda = 2\pi r$, substitute for wavelength using de Broglie's relation $\lambda = \dfrac{h}{mv}$:

$n\dfrac{h}{mv} = 2\pi r$, which rearranges to $nh = 2\pi mvr$. Finally, dividing both sides by 2π yields

$mvr = \dfrac{nh}{2\pi}$, which is Bohr's postulate that the angular momentum, mvr, is a quantized multiple of

$h/2\pi$.

9.37. $\lambda = \dfrac{h}{mv} = 1.00 \times 10^{-10} \text{ m} = \dfrac{(6.626 \times 10^{-34} \text{ Js})}{(9.109 \times 10^{-31} \text{ kg})(v)}$ $v = 7.27 \times 10^6$ m/s for an electron.

$\lambda = \dfrac{h}{mv} = 1.00 \times 10^{-10} \text{ m} = \dfrac{(6.626 \times 10^{-34} \text{ Js})}{(1.672 \times 10^{-27} \text{ kg})(v)}$ $v = 3.96 \times 10^3$ m/s for a proton.

CHAPTER 10. INTRODUCTION TO QUANTUM MECHANICS

10.1. From the text, the following statements were given as postulates. The state of a system can be described by an expression called a wavefunction; all possible information about the observables of the system are contained in the wavefunction. For every physical observable, there exists a corresponding operator, and the value of the observable can be determined by an eigenvalue equation involving the operator and the wavefunction. Allowable wavefunctions must satisfy the Schrodinger equation. The average value of an observable can be determined from the appropriate operator and wavefunction using the expression given by equation 10.13. Other sources may list postulates differently; your answer may vary.

10.3. (a) Yes, the function is an acceptable wavefunction. (b) No, the function is not acceptable because not finite over the range given. Notice that over a certain range, the function is imaginary. This does not disqualify it as a possible wavefunction! (c) No, the function is not acceptable because it is not continuous. (d) Yes, the function is an acceptable wavefunction. (e) No, the function is not acceptable because it is not bounded. (f) Yes, the function is an acceptable wavefunction. (g) No, the function is not acceptable because it is not single-valued. (h) No, the function is not acceptable because it is not continuous. (g) No, the function is not acceptable because it is not single-valued.

10.5. (a) Evaluation of the operation yields the value 6. (b) Evaluation of the operation yields the value 9. (c) Evaluation of the operation yields the function $12x^2 - 7 - \dfrac{7}{x^2}$.

10.7. (a) $\hat{P}_x(4,5,6) = (-4,5,6)$ (b) $\hat{P}_y\hat{P}_z(0,-4,-1) = (0,4,1)$

(c) $\hat{P}_x\hat{P}_x(5,0,0) = (5,0,0)$ (the two operations cancel each other out)

(d) $\hat{P}_y\hat{P}_x(\pi,\pi/2,0) = (-\pi,-\pi/2,0)$

(e) Yes, $\hat{P}_x\hat{P}_y$ should equal $\hat{P}_y\hat{P}_x$ for any set of coordinates, because the x coordinate and the y coordinate are independent of each other.

10.9. When multiplying a function by a constant, the result is simply the original function times a constant. Since this is what's required by an eigenvalue equation, multiplying a function by a constant always yields an eigenvalue and the original function.

10.11. In terms of classical mechanics, $K = \frac{1}{2} mv^2$, or in terms of linear momentum $p = mv$, $K = \dfrac{p^2}{2m}$. Substituting for the definition of the linear momentum operator $\hat{p} = -i\hbar \dfrac{\partial}{\partial x}$:

$\hat{K} = \dfrac{1}{2m}\left(-i\hbar \dfrac{\partial}{\partial x}\right)^2 = \dfrac{1}{2m}\left(-\hbar^2 \dfrac{\partial^2}{\partial x^2}\right) = -\dfrac{\hbar^2}{2m}\dfrac{\partial^2}{\partial x^2}$, which is the one-dimensional kinetic energy operator.

10.13. $-i\hbar\dfrac{\partial}{\partial\phi}\left(\dfrac{1}{\sqrt{2\pi}}e^{im\phi}\right) = -i\hbar(im)\left(\dfrac{1}{\sqrt{2\pi}}e^{im\phi}\right) = m\hbar\left(\dfrac{1}{\sqrt{2\pi}}e^{im\phi}\right)$. Since the original function is returned multiplied by a constant, the function is an eigenfunction and the value of the angular momentum is $m\hbar$.

10.15. $1.23m_e = 1.23(9.109\times10^{-31}\text{ kg}) = 1.12\times10^{-30}\text{ kg}$

$\Delta x\left(1.12\times10^{-30}\text{ kg}\cdot\dfrac{10,000\text{ m}}{\text{s}}\right) \geq \dfrac{6.626\times10^{-34}\text{ J}\cdot\text{s}}{4\pi}$ $\qquad\Delta x \geq 4.71\times10^{-9}\text{ m}$.

10.17. First, we need to convert the 1.00 cm^{-1} to energy:

$\dfrac{1}{1.00\text{ cm}^{-1}}\left(\dfrac{1\text{ m}}{100\text{ cm}}\right) = 0.01\text{ m} = \lambda$ $\qquad v = \dfrac{c}{\lambda} = \dfrac{3.00\times10^{8}\text{ m/s}}{0.01\text{ m}} = 3.00\times10^{10}\text{ s}^{-1}$

Now calculating the equivalent ΔE: $\Delta E = hv = (6.626\times10^{-34}\text{ J}\cdot\text{s})(3.00\times10^{10}\text{ s}^{-1})$

$\Delta E = 1.99\times10^{-23}\text{ J}$ Calculating the Δt from the expression in the exercise:

$(1.99\times10^{-23}\text{ J})\Delta t = \dfrac{6.626\times10^{-34}\text{ J}\cdot\text{s}}{4\pi}$ $\qquad\Delta t = 2.65\times10^{-12}\text{ s}$.

This is just under 3 picoseconds.

10.19. (a) To normalize: $N^2\displaystyle\int_0^{2\pi}(e^{im\phi})^*e^{im\phi}\,d\phi = 1$. $N^2\displaystyle\int_0^{2\pi}e^{-im\phi}e^{im\phi}\,d\phi = 1$

$N^2\displaystyle\int_0^{2\pi}1\cdot d\phi = 1$ $\qquad N^2[\phi]_0^{2\pi} = 1$ $\qquad N^2(2\pi - 0) = 1$ $\qquad N^2 = \dfrac{1}{2\pi}$ $\qquad N = \dfrac{1}{\sqrt{2\pi}}$

Therefore, the normalized wavefunction is $\dfrac{1}{\sqrt{2\pi}}e^{im\phi}$.

(b) $P = \displaystyle\int_0^{2\pi/3}\left(\dfrac{1}{\sqrt{2\pi}}e^{im\phi}\right)^*\dfrac{1}{\sqrt{2\pi}}e^{im\phi}\,d\phi = \dfrac{1}{2\pi}\int_2^{2\pi/3}e^{-im\phi}\cdot e^{im\phi}\,d\phi = \dfrac{1}{2\pi}\int_2^{2\pi/3}d\phi$

$= \dfrac{1}{2\pi}[\phi]_0^{2\pi/3} = \dfrac{1}{2\pi}\left(\dfrac{2\pi}{3} - 0\right) = \dfrac{1}{3}$. The probability is 1/3. This makes sense because the region of interest is 1/3 of the ring. The probability does not depend on m, so any value of m would have the same value of probability over this interval.

10.21. First, we need to normalize the function:

$N^2\displaystyle\int_0^{a}(kx)^*\,kx\,dx = 1$. $\qquad N^2k^2\displaystyle\int_0^{a}x^2\,dx = 1$ $\qquad N^2k^2\left[\dfrac{1}{3}x^3\right]_0^{a} = 1$

$N^2k^2\left(\dfrac{1}{3}a^3 - 0\right) = 1$ $\qquad N^2k^2a^3 = 3$ $\qquad N^2 = \dfrac{3}{k^2a^3}$ $\qquad N = \dfrac{\sqrt{3}}{ka^{3/2}}$

Therefore the normalized wavefunction is $\Psi = \dfrac{\sqrt{3}}{ka^{3/2}}kx = \dfrac{\sqrt{3}}{a^{3/2}}x$. Again, note how the k cancels.

For the first third of the interval:

$$P = \int_0^{a/3} \left(\frac{\sqrt{3}}{a^{3/2}} x \right)^* \frac{\sqrt{3}}{a^{3/2}} x \, dx = \frac{3}{a^3} \int_0^{a/3} x^2 \, dx = \frac{3}{a^3} \cdot \frac{1}{3} x^3 \Big|_0^{a/3} = \frac{3}{a^3} \left(\frac{1}{3} \frac{a^3}{27} - 0 \right) = \frac{1}{27}.$$

For the last third of the interval:

$$P = \int_{2a/3}^{a} \left(\frac{\sqrt{3}}{a^{3/2}} x \right)^* \frac{\sqrt{3}}{a^{3/2}} x \, dx = \frac{3}{a^3} \int_{2a/3}^{a} x^2 \, dx = \frac{3}{a^3} \cdot \frac{1}{3} x^3 \Big|_{2a/3}^{a} = \frac{3}{a^3} \left(\frac{1}{3} a^3 - \frac{1}{3} \frac{8a^3}{27} \right) = \frac{19}{27}.$$

In this case, the particle has a higher probability of being in the last third of the interval, and the probability distribution is not equal like it is in the previous problem.

10.23. (a) $N^2 \int_0^1 (x^2)^* x^2 \, dx = 1$ $\qquad N^2 \int_0^1 x^4 \, dx = 1$ $\qquad N^2 \left[\frac{1}{5} x^5 \right]_0^1 = 1$

$N^2(1/5 - 0) = 1$ $\qquad N^2 = 5$ $\qquad N = \sqrt{5}$ $\qquad$ Therefore, the normalized wavefunction is $\Psi = \sqrt{5} x^2$.

(b) $N^2 \int_5^6 (1/x)^* (1/x) \, dx = 1$ $\qquad N^2 \int_5^6 (1/x)^2 \, dx = 1$ $\qquad N^2 \left[-\frac{1}{x} \right]_5^6$

$N^2 \left(-\frac{1}{6} - \left(-\frac{1}{5} \right) \right) = 1$ $\qquad N^2 \left(\frac{1}{30} \right) = 1$ $\qquad N = \sqrt{30}$ Therefore, the normalized

wavefunction is $\Psi = \dfrac{\sqrt{30}}{x}$.

(c) $N^2 \int_{-\pi/2}^{\pi/2} (\cos x)^* (\cos x) \, dx = 1$ $\qquad N^2 \int_{-\pi/2}^{\pi/2} \cos^2 x \, dx = 1$ Using the integral table in the

appendix:

$N^2 \left[\frac{x}{2} + \frac{1}{4} \sin 2x \right]_{-\pi/2}^{\pi/2} = 1$ $\qquad N^2 \left[\left(\frac{\pi/2}{2} + \frac{1}{4} \sin 2(\pi/2) \right) - \left(\frac{-\pi/2}{2} + \frac{1}{4} \sin 2(-\pi/2) \right) \right] = 1$

$N^2 \left[\frac{\pi}{4} + \frac{\pi}{4} + \frac{1}{4}(0 - 0) \right] = 1$ $\qquad N^2 \left(\frac{\pi}{2} \right) = 1$ $\qquad N^2 = \frac{2}{\pi}$ $\qquad N = \sqrt{\frac{2}{\pi}}$. Therefore,

the normalized wavefunction is $\Psi = \sqrt{\dfrac{2}{\pi}} \cos x$.

(d) $N^2 \int_0^{\infty} (e^{-r/a})^* (e^{-r/a}) 4\pi r^2 \, dr = 1$ $\qquad 4\pi N^2 \int_0^{\infty} r^2 e^{-2r/a} \, dr = 1$ Using the integral table in

the appendix: $4\pi N^2 \left[\frac{2!}{(2/a)^3} \right] = 1$ $\qquad 8\pi N^2 a^3 = 1$ $\qquad N = \dfrac{1}{\sqrt{8\pi a^3}}$ Therefore, the

normalized wavefunction is $\Psi = \dfrac{1}{\sqrt{8\pi a^3}} e^{-r/a}$.

(e) The tactic for this wavefunction is to recognize that since the exponent has the variable squared, the wavefunction in the interval from $-\infty$ to 0 has the same as it has in the interval 0 to

44

∞. Therefore, we can use the interval 0 to ∞ and multiply the value of the integral by 2. Therefore, we have

$$2N^2 \int_0^\infty (e^{-r^2/a})^* (e^{-r^2/a}) 4\pi r^2 \, dr = 1 \qquad 8\pi N^2 \int_0^\infty r^2 e^{-2r^2/a} \, dr = 1 \quad \text{Using the integral table in}$$

the appendix: $8\pi N^2 \left(\dfrac{1}{2^2 \cdot (2/a)^1} \cdot \sqrt{\dfrac{\pi}{(2/a)}} \right) = 1 \qquad \pi N^2 a \sqrt{\dfrac{\pi a}{2}} = 1$

$N = \left(\dfrac{2}{\pi^3 a^3} \right)^{1/4}$ Therefore, the normalized wavefunction is $\Psi = \left(\dfrac{2}{\pi^3 a^3} \right)^{1/4} e^{r^2/a}$.

10.25. The Schrödinger equation contains a specific expression for the kinetic energy because kinetic energy has a specific formula in classical mechanics ($\dfrac{1}{2}mv^2$, or $\dfrac{p^2}{2m}$ - the two are equivalent) which can be transferred into a kinetic energy operator. However, there is no single expression for the potential energy of a system. The expression of the potential energy depends on the definition of the system. Therefore, in the Schrödinger equation, there is only a general expression for the potential energy operator.

10.27. $-\dfrac{\hbar^2}{2m} \dfrac{\partial^2}{\partial x^2}(k) + 0 \cdot k = Ek$ Since the derivative of a constant is zero, we have

$0 + 0 = E \cdot k$, Therefore, E must be 0.

10.29. The only property of Hermitian operators described in the text is that Hermitian operators have real (i.e. non-imaginary) eigenvalues. While not a strict proof, we can argue that if the Hamiltonian operator yields energy as an observable, and since we know that energy is a real quantity, we submit that the Hamiltonian must produce real eigenvalues and so be a Hermitian operator. (There are other mathematical requirements for Hermitian operators, but this text doesn't cover them.)

10.31. To normalize: $N^2 \int_0^{2\pi} (e^{iKx})^* e^{iKx} \, dx = 1$. $N^2 \int_0^{2\pi} e^{-iKx} e^{iKx} \, dx = 1$

$N^2 \int_0^{2\pi} 1 \cdot dx = 1 \qquad N^2 [x]_0^{2\pi} = 1 \qquad N^2(2\pi - 0) = 1 \qquad N^2 = \dfrac{1}{2\pi} \qquad N = \dfrac{1}{\sqrt{2\pi}}$

Therefore, the normalized wavefunction is $\dfrac{1}{\sqrt{2\pi}} e^{iKx}$. Now, substituting this into the

Schrödinger equation: $-\dfrac{\hbar^2}{2m} \dfrac{\partial^2}{\partial x^2} \left(\dfrac{1}{\sqrt{2\pi}} e^{iKx} \right) = (iK)^2 \left(-\dfrac{\hbar^2}{2m} \right) \left(\dfrac{1}{\sqrt{2\pi}} e^{iKx} \right) = \dfrac{\hbar^2 K^2}{2m} \left(\dfrac{1}{\sqrt{2\pi}} e^{iKx} \right)$.

Therefore it is an eigenfunction with an eigenvalue of $\dfrac{\hbar^2 K^2}{2m}$. The eigenvalue doesn't change with inclusion of a normalization constant.

10.33. First, we should convert the wavelength of the photon into an equivalent value in joules: $\lambda = 2170$ Å $= 2.17\times10^{-7}$ m. Therefore, $v = \dfrac{c}{\lambda} = \dfrac{3.00\times10^{8} \text{ m/s}}{2.17\times10^{-7} \text{ m}} = 1.38\times10^{15}$ s^{-1}. The energy of a photon of this frequency is $E = hv = (6.626\times10^{-34} \text{ J·s})(1.38\times10^{15} \text{ s}^{-1}) = 9.14\times10^{-19}$ J. Now we use the expression for the energies of the particle-in-a-box wavefunctions:

9.14×10^{-19} J $= E(n = 3) - E(n = 2) = \dfrac{(3^2 - 2^2)(6.626\times10^{-34} \text{ J·s})^2}{8(9.109\times10^{-31} \text{ kg})a^2}$ We have all of the

information except the length of the 'box', a. Solve algebraically: $a = 5.74\times10^{-10}$ m $= 5.74$ Å.

10.35. The drawings are left to the student. See Figures 10.6 and 10.7 in the text.

10.37. For $n = 1$: $P = \displaystyle\int_{0.495a}^{0.505a} \left(\sqrt{\dfrac{2}{a}}\sin\dfrac{\pi x}{a}\right)^{*}\left(\sqrt{\dfrac{2}{a}}\sin\dfrac{\pi x}{a}\right)dx = \dfrac{2}{a}\displaystyle\int_{0.495a}^{0.505a}\sin^2\dfrac{\pi x}{a}dx$

$\dfrac{2}{a}\left[\dfrac{x}{2} - \dfrac{1}{4(\pi/a)}\sin\dfrac{2\pi x}{a}\right]_{0.495a}^{0.505a}$

$= \dfrac{2}{a}\left[\left(\dfrac{0.505a}{2} - \dfrac{a}{4\pi}\sin 2\pi(0.505)\right) - \left(\dfrac{0.495a}{2} - \dfrac{a}{4\pi}\sin 2\pi(0.495)\right)\right]$

$P = 2(0.005 + 0.00250 + 0.00250) = 0.02$

For $n = 2$: $P = \displaystyle\int_{0.495a}^{0.505a} \left(\sqrt{\dfrac{2}{a}}\sin\dfrac{2\pi x}{a}\right)^{*}\left(\sqrt{\dfrac{2}{a}}\sin\dfrac{2\pi x}{a}\right)dx = \dfrac{2}{a}\displaystyle\int_{0.495a}^{0.505a}\sin^2\dfrac{2\pi x}{a}dx$

$\dfrac{2}{a}\left[\dfrac{x}{2} - \dfrac{1}{4(2\pi/a)}\sin\dfrac{4\pi x}{a}\right]_{0.495a}^{0.505a}$

$= \dfrac{2}{a}\left[\left(\dfrac{0.505a}{2} - \dfrac{a}{8\pi}\sin 4\pi(0.505)\right) - \left(\dfrac{0.495a}{2} - \dfrac{a}{8\pi}\sin 4\pi(0.495)\right)\right]$

$P = 2(0.005 - 0.002498 - 0.002498) = 0.000008$

For $n = 3$: $P = \displaystyle\int_{0.495a}^{0.505a} \left(\sqrt{\dfrac{2}{a}}\sin\dfrac{3\pi x}{a}\right)^{*}\left(\sqrt{\dfrac{2}{a}}\sin\dfrac{3\pi x}{a}\right)dx = \dfrac{2}{a}\displaystyle\int_{0.495a}^{0.505a}\sin^2\dfrac{3\pi x}{a}dx$

$\dfrac{2}{a}\left[\dfrac{x}{2} - \dfrac{1}{4(3\pi/a)}\sin\dfrac{6\pi x}{a}\right]_{0.495a}^{0.505a}$

$= \dfrac{2}{a}\left[\left(\dfrac{0.505a}{2} - \dfrac{a}{12\pi}\sin 6\pi(0.505)\right) - \left(\dfrac{0.495a}{2} - \dfrac{a}{12\pi}\sin 6\pi(0.495)\right)\right]$

$P = 2(0.005 + 0.002496 + 0.002496) = 0.01998$

For $n = 4$: $P = \displaystyle\int_{0.495a}^{0.505a} \left(\sqrt{\dfrac{2}{a}}\sin\dfrac{4\pi x}{a}\right)^{*}\left(\sqrt{\dfrac{2}{a}}\sin\dfrac{4\pi x}{a}\right)dx = \dfrac{2}{a}\displaystyle\int_{0.495a}^{0.505a}\sin^2\dfrac{4\pi x}{a}dx$

$$\frac{2}{a}\left[\frac{x}{2}-\frac{1}{4(4\pi/a)}\sin\frac{8\pi x}{a}\right]_{0.495a}^{0.505a}$$

$$=\frac{2}{a}\left[\left(\frac{0.505a}{2}-\frac{a}{16\pi}\sin 8\pi(0.505)\right)-\left(\frac{0.495a}{2}-\frac{a}{16\pi}\sin 8\pi(0.495)\right)\right]$$

$P = 2(0.005 - 0.002493 - 0.002493) = 0.00028$

The probabilities are relatively high for the first and third wavefunctions (as they should be, as the middle of the box is the point where the wavefunctions have the highest magnitude), and relatively low for the second and fourth wavefunctions (as they should be, as the middle of the box is where these wavefunctions have a node).

10.39. At high quantum numbers, the positive peaks of the probability waves effectively blend together, mimicking a straight line of constant probability across the box. This is what would be expected for a classical particle bouncing back and forth between two walls of a box, which is consistent with the correspondence principle.

10.41. The wavefunctions $\Psi = \sqrt{\frac{2}{a}}\sin\frac{n\pi x}{a}$ can be very easily shown to not be eigenfunctions of the position operator. The position operator $\hat{x}$ is defined as "$x\cdot$"; that is, multiplication by the x variable. This generates a new function, the sine function times the x variable. Since this is a new function and not the original function times a constant, the wavefunctions are not eigenfunctions.

10.43. $\langle p_x \rangle = \int_0^a \left(\sqrt{\frac{2}{a}}\sin\frac{\pi x}{a}\right)^* \cdot -i\hbar\frac{\partial}{\partial x}\sqrt{\frac{2}{a}}\sin\frac{\pi x}{a}\,dx = -i\hbar\frac{2}{a}\int_0^a \sin\frac{\pi x}{a}\frac{\partial}{\partial x}\sin\frac{\pi x}{a}\,dx$

$= -i\hbar\frac{2}{a}\frac{\pi}{a}\int_0^a \sin\frac{\pi x}{a}\cos\frac{\pi x}{a}\,dx$

$= -i\hbar\frac{2\pi}{a^2}\left[\frac{1}{\pi/a}\sin^2\frac{\pi x}{a}\right]_0^a = -i\hbar\frac{2\pi}{a^2}\left[\frac{a}{\pi}(\sin\pi - \sin 0)\right] = -i\hbar\frac{2\pi}{a^2}\left[\frac{a}{\pi}(0-0)\right] = 0$

10.45. A normalization constant is not going to affect the value of the eigenvalue, so let us simply evaluate the operator/wavefunction combination without normalizing:

$-i\hbar\frac{\partial}{\partial\phi}e^{3i\phi} = -i\hbar(3i)e^{3i\phi} = 3\hbar e^{3i\phi}$. Therefore, the eigenvalue is $3\hbar$, that being the value of the angular momentum. To determine the average value, set up the expression as follows, using N as the normalization constant:

$\langle p_\phi \rangle = \int_0^{2\phi}\left(Ne^{3i\phi}\right)^* \cdot -i\hbar\frac{\partial}{\partial\phi}Ne^{3i\phi}\,d\phi = -i\hbar\cdot 3i\int_0^{2\pi}Ne^{-3i\phi}\cdot Ne^{3i\phi}\,d\phi$. If the wavefunction is properly normalized, the value of the integral is simply 1. Therefore, we have $\langle p_\phi \rangle = 3\hbar$, which is the same value as the eigenvalue for angular momentum, as it should be.

10.47. Defining $\dfrac{1}{X}\dfrac{d^2}{dx^2}X$ as $-\dfrac{2mE}{\hbar^2}$ allows us to ultimately rewrite the expression into a form that mimics a one-dimensional Schrödinger equation. If we had defined the second derivative as simply E, the connection to the one-dimensional Schrödinger equation would not have been as obvious. (See also the next exercise.)

10.49. $-\dfrac{\hbar^2}{2m}\left(\dfrac{\partial^2}{\partial x^2}+\dfrac{\partial^2}{\partial y^2}+\dfrac{\partial^2}{\partial z^2}\right)\left[\sqrt{\dfrac{8}{abc}}\sin\dfrac{n_x\pi x}{a}\cdot\sin\dfrac{n_y\pi y}{b}\cdot\sin\dfrac{n_z\pi z}{c}\right]$

$=-\dfrac{\hbar^2}{2m}\left[\sqrt{\dfrac{8}{abc}}\left(\dfrac{\partial^2}{\partial x^2}\sin\dfrac{n_x\pi x}{a}\right)\cdot\sin\dfrac{n_y\pi y}{b}\cdot\sin\dfrac{n_z\pi z}{c}+\sqrt{\dfrac{8}{abc}}\sin\dfrac{n_x\pi x}{a}\cdot\left(\dfrac{\partial^2}{\partial y^2}\sin\dfrac{n_y\pi y}{b}\right)\cdot\sin\dfrac{n_z\pi z}{c}\right.$

$\left.+\sqrt{\dfrac{8}{abc}}\sin\dfrac{n_x\pi x}{a}\cdot\sin\dfrac{n_y\pi y}{b}\left(\dfrac{\partial^2}{\partial z^2}\sin\dfrac{n_z\pi z}{c}\right)\cdot\right]$

$=-\dfrac{\hbar^2}{2m}\left[-\dfrac{n_x^2\pi^2}{a^2}\sqrt{\dfrac{8}{abc}}\sin\dfrac{n_x\pi x}{a}\cdot\sin\dfrac{n_y\pi y}{b}\cdot\sin\dfrac{n_z\pi z}{c}-\dfrac{n_y^2\pi^2}{b^2}\sqrt{\dfrac{8}{abc}}\sin\dfrac{n_x\pi x}{a}\cdot\sin\dfrac{n_y\pi y}{b}\cdot\sin\dfrac{n_z\pi z}{c}\right.$

$\left.-\dfrac{n_z^2\pi^2}{c^2}\sqrt{\dfrac{8}{abc}}\sin\dfrac{n_x\pi x}{a}\cdot\sin\dfrac{n_y\pi y}{b}\cdot\sin\dfrac{n_z\pi z}{c}\right]$

$=\dfrac{\hbar^2}{2m}\left[\dfrac{n_x^2\pi^2}{a^2}+\dfrac{n_y^2\pi^2}{b^2}+\dfrac{n_z^2\pi^2}{c^2}\right]\Psi=\dfrac{h^2}{8m}\left[\dfrac{n_x^2}{a^2}+\dfrac{n_y^2}{b^2}+\dfrac{n_z^2}{c^2}\right]\Psi$, where the energy eigenvalue is

$\dfrac{h^2}{8m}\left[\dfrac{n_x^2}{a^2}+\dfrac{n_y^2}{b^2}+\dfrac{n_z^2}{c^2}\right]$. This verifies that the wavefunction is an eigenfunction of the three-dimensional Schrödinger equation.

10.51. Figure 10.13 can be consulted to answer this exercise. The first set of quantum numbers for which degeneracy occurs is (1,1,2), whose energy is the same as the sets (1,2,1) and (2,1,1). The first appearance of degeneracy for sets of different quantum numbers (as opposed to simply the rearrangement of the same quantum numbers) is (5,1,1) and its iterations (1,5,1) and (1,1,5) with the set (3,3,3).

10.53. For the two-dimensional box: $\hat{H}=-\dfrac{\hbar^2}{2m}\left(\dfrac{\partial^2}{\partial x^2}+\dfrac{\partial^2}{\partial y^2}\right)$, the wavefunctions are

$\Psi=\sqrt{\dfrac{4}{ab}}\sin\dfrac{n_x\pi x}{a}\sin\dfrac{n_y\pi y}{b}$, and the quantized energies are $\dfrac{h^2}{8m}\left[\dfrac{n_x^2}{a^2}+\dfrac{n_y^2}{b^2}\right]$.

10.55. We must evaluate the following integral:

$\langle x^2\rangle=\iiint\left(\sqrt{\dfrac{8}{abc}}\sin\dfrac{\pi x}{a}\cdot\sin\dfrac{\pi y}{b}\cdot\sin\dfrac{\pi z}{c}\right)^*\cdot x^2\cdot\left(\sqrt{\dfrac{8}{abc}}\sin\dfrac{\pi x}{a}\cdot\sin\dfrac{\pi y}{b}\cdot\sin\dfrac{\pi z}{c}\right)dx\,dy\,dz$

We can simplify this by separating the wavefunction (including the normalization constant) into three parts, one part for x, one part for y, and one part for z:

$$\langle x^2 \rangle = \int_0^a \left(\sqrt{\frac{2}{a}} \sin \frac{\pi x}{a} \right)^* \cdot x^2 \cdot \left(\sqrt{\frac{2}{a}} \sin \frac{\pi x}{a} \right) dx$$

$$\times \int_0^b \left(\sqrt{\frac{2}{b}} \sin \frac{\pi y}{b} \right)^* \left(\sqrt{\frac{2}{b}} \sin \frac{\pi y}{b} \right) dy \times \int_0^c \left(\sqrt{\frac{2}{c}} \sin \frac{\pi z}{c} \right)^* \left(\sqrt{\frac{2}{c}} \sin \frac{\pi z}{c} \right) dz$$

The reason we do this is so we can now recognize that the second and third integrals represent normalized one-dimensional wavefunctions, so the integrals are simply 1. Therefore, we get

$$\langle x^2 \rangle = \int_0^a \left(\sqrt{\frac{2}{a}} \sin \frac{\pi x}{a} \right)^* x^2 \sqrt{\frac{2}{a}} \sin \frac{\pi x}{a} dx = \frac{2}{a} \int_0^a x^2 \sin^2 \frac{\pi x}{a} dx .$$ Using the integral table in the

appendix:

$$= \frac{2}{a} \left[\frac{x^3}{6} - \left(\frac{x^2}{4(\pi/a)} - \frac{1}{8(\pi/a)^3} \right) \sin \frac{2\pi x}{a} - \frac{x}{4(\pi/a)^2} \cos \frac{2\pi x}{a} \right]_0^a$$

$$= \frac{2}{a} \left[\left(\frac{a^3}{6} - 0 - \frac{a^3}{4\pi^2} \right) - \left(0 - 0 - \frac{a^3}{4\pi^2} \right) \right] = \frac{a^2}{3}$$

Finally, rather than repeat everything, by extension we can deduce that $\langle y^2 \rangle = \frac{b^2}{3}$ and

$$\langle z^2 \rangle = \frac{c^2}{3} .$$

10.57. The first integral is set up as $\int_0^a \left(\sqrt{\frac{2}{a}} \sin \frac{\pi x}{a} \right)^* \sqrt{\frac{2}{a}} \sin \frac{2\pi x}{a} dx$. Using the appendix:

$$= \frac{2}{a} \left[\frac{\sin(\pi/a - 2\pi/a)x}{2(\pi/a - 2\pi/a)} + \frac{\sin(\pi/a + 2\pi/a)x}{2(\pi/a + 2\pi/a)} \right]_0^a = \frac{2}{a} \left[\frac{\sin(-\pi)}{-2\pi/a} + \frac{\sin(+\pi)}{-2\pi/a} \right] = \frac{2}{a}(0+0) = 0$$

The reverse order of wavefunctions gives us $\int_0^a \left(\sqrt{\frac{2}{a}} \sin \frac{2\pi x}{a} \right)^* \sqrt{\frac{2}{a}} \sin \frac{\pi x}{a} dx$. Solving:

$$= \frac{2}{a} \left[\frac{\sin(2\pi/a - \pi/a)x}{2(2\pi/a - \pi/a)} + \frac{\sin(2\pi/a + \pi/a)x}{2(2\pi/a + \pi/a)} \right]_0^a = \frac{2}{a} \left[\frac{\sin(\pi)}{2\pi/a} + \frac{\sin(-\pi)}{2\pi/a} \right] = \frac{2}{a}(0+0) = 0$$

Therefore, order doesn't matter when showing that any two wavefunctions are orthogonal.

10.59. $\hat{H} e^{-iEt/\hbar} \Psi(x) = i\hbar \dfrac{\partial \left(e^{-iEt/\hbar} \Psi(x) \right)}{\partial t}$ is the expression we need to show is satisfied. On the

left side, we recognize that the Hamiltonian operator does not include time, so the exponential function can be moved outside of the operator. On the right side, the spatial part of the wavefunction, $\Psi(x)$, does not depend on time, so it can be removed from the derivative. We get:

$e^{-iEt/\hbar} \cdot \hat{H}\Psi(x) = i\hbar\Psi(x)\dfrac{\partial(e^{-iEt/\hbar})}{\partial t}$. Evaluating the derivative: $\dfrac{\partial(e^{-iEt/\hbar})}{\partial t} = e^{-iEt/\hbar} \cdot -\dfrac{iE}{\hbar}$.

Substituting:

$e^{-iEt/\hbar} \cdot \hat{H}\Psi(x) = i\hbar \cdot -\dfrac{iE}{\hbar} \cdot e^{-iEt/\hbar} \cdot \Psi(x)$. The two exponential functions cancel. On the right side of the equation, the $\hbar$ terms cancel, as do the two i's with the negative sign. What is left is $\hat{H}\Psi(x) = E\Psi(x)$, which is the time-independent Schrödinger equation. Since we accept this equation as a postulate, we affirm that the original equation does in fact satisfy the time-dependent Schrödinger equation.

10.61. $|\Psi(x,t)|^2 = (\Psi(x,t))^* \Psi(x,t) = [e^{-iEt/\hbar} \cdot \Psi(x)]^* e^{-iEt/\hbar} \cdot \Psi(x) = e^{+iEt/\hbar} \cdot \Psi(x)^* \cdot e^{-iEt/\hbar} \cdot \Psi(x)$

$= \Psi(x)^* \cdot \Psi(x) = |\Psi(x)|^2$. Thus, the square magnitudes of the time-dependent and the time-independent wavefunctions are the same.

CHAPTER 11. QUANTUM MECHANICS: MODEL SYSTEMS AND THE HYDROGEN ATOM

11.1. $3.558 \dfrac{\text{mdyn}}{\text{A}} \times \dfrac{1\,\text{dyn}}{1000\,\text{mdyn}} \times \dfrac{10^{10}\,\text{A}}{1\,\text{m}} \times \dfrac{1\,\text{N}}{10^5\,\text{dyn}} = 355.8 \dfrac{\text{N}}{\text{m}}$.

11.3. Objects in a gravitational field are experiencing a force that is directly proportional to their height above their equilibrium heights. Thus, they satisfy the fundamental mathematical requirement for a Hooke's-law type of oscillator.

11.5. Starting with $\left[-\dfrac{\hbar^2}{2m}\dfrac{d^2}{dx^2} + 2\pi^2 \nu^2 m x^2 \right]\Psi = E\Psi$, substitute the definition to α to get

$\left[-\dfrac{\hbar^2}{2m}\dfrac{d^2}{dx^2} + \dfrac{\alpha^2 x^2 \hbar^2}{2m} \right]\Psi = E\Psi$. Dividing both sides of the equation by $-\dfrac{\hbar^2}{2m}$ yields

$\left[\dfrac{d^2}{dx^2} - \alpha^2 x^2 \right]\Psi = -\dfrac{2mE}{\hbar^2}\Psi$. Finally, bringing all terms to one side of the equation and

factoring out the Ψ from two terms, we get $\dfrac{d^2\Psi}{dx^2} + \left(\dfrac{2mE}{\hbar^2} - \alpha^2 x^2 \right)\Psi = 0$, which is our ultimately

result.

11.7. Starting with $\alpha + 2\alpha n - \dfrac{2mE}{\hbar^2} = 0$, first let us multiply all terms by $\dfrac{\hbar^2}{2m}$. We get

$\dfrac{\alpha \hbar^2}{2m} + \dfrac{\alpha n \hbar^2}{m} - E = 0$, which rearranges to $E = \dfrac{\alpha \hbar^2}{2m} + \dfrac{\alpha n \hbar^2}{m}$. Now we substitute for the

definition of α: $E = \dfrac{2\pi \nu m \hbar^2}{2m\hbar} + \dfrac{2\pi \nu m n \hbar^2}{m\hbar}$. Recalling that $\hbar = \dfrac{h}{2\pi}$, we can substitute and cancel

out the 2π terms. We also cancel mass out of both fractions. We are left with $E = \dfrac{\nu h}{2} + n\nu h$.

Factoring out the $h\nu$ from both terms, we are left with $E = \left(n + \dfrac{1}{2} \right)h\nu$, which is the required

equation.

11.9. (a) $\Delta E = h\nu$ $\qquad\qquad \Delta E = (6.626\times10^{-34}\,\text{J·s})(1.00\,\text{s}^{-1}) = 6.63\times10^{-34}\,\text{J}$.

(b) $\lambda = \dfrac{c}{\nu} = \dfrac{3.00\times10^8\,\text{m/s}}{1.00\,\text{s}^{-1}} = 3.00\times10^8\,\text{m}$ $\qquad$ **(c)** This is in the (very-long-wavelength) radio

wave region of the electromagnetic spectrum. **(d)** Such a long, low-energy radio wave would have been undetectable by early-20th-century technology (and may still be undetectable today!).

11.11. Assuming that the energy change between each level is the same, the overall energy change for the process will be four times the energy interval between adjacent levels, or $\Delta E = 4h\nu$. Now we need to convert our vibrational frequency to a true frequency.

$$\tilde{\nu} = 3650 \text{ cm}^{-1} = \frac{1}{\lambda} \qquad \lambda = 2.74 \times 10^{-4} \text{ cm} = 2.74 \times 10^{-6} \text{ m} \quad \text{Now, we calculate the frequency of}$$

this light: $\nu = \dfrac{c}{\lambda} = \dfrac{3.00 \times 10^{-8} \text{ m/s}}{2.74 \times 10^{-6} \text{ m}} = 1.09 \times 10^{14} \text{ s}^{-1}$. Now we can calculate the energy:

$\Delta E = 4(6.626 \times 10^{-34} \text{ J} \cdot \text{s})(1.09 \times 10^{14} \text{ s}^{-1}) = 2.89 \times 10^{-19}$ J. For one photon to have that amount of

energy, it must have a frequency of $\dfrac{2.89 \times 10^{-19} \text{ J}}{6.626 \times 10^{-34} \text{ J} \cdot \text{s}} = 4.36 \times 10^{14} \text{ s}^{-1}$ (which is four times the

original frequency). The wavelength of a photon having that frequency is

$$\lambda = \frac{c}{\nu} = \frac{3.00 \times 10^{8} \text{ m/s}}{4.36 \times 10^{14} \text{ s}^{-1}} = 6.88 \times 10^{-7} \text{ m}.$$

11.13. Using equation 11.19, we find that Ψ_1 has the form $\left(\dfrac{\alpha}{\pi}\right)^{1/4} \left(\dfrac{1}{2}\right)^{1/2} (2\alpha^{1/2} x) e^{-\alpha x^2/2}$.

Substituting into the Schrödinger equation: $-\dfrac{\hbar^2}{2m} \dfrac{\partial^2}{\partial x^2} N(2\alpha^{1/2} x) e^{-\alpha x^2/2} + \dfrac{1}{2} kx^2 N(2\alpha^{1/2} x) e^{-\alpha x^2/2}$

(where we are using N to represent the collection of constants at the beginning of the wavefunction, for clarity). First, let us evaluate the second derivative of the wavefunction.

The first derivative is: $\dfrac{\partial}{\partial x}(2\alpha^{1/2} x) e^{-\alpha x^2/2} = 2\alpha^{1/2} x e^{-\alpha x^2/2} \cdot (-\alpha x) + 2\alpha^{1/2} e^{-\alpha x^2/2}$

The second derivative is:

$\dfrac{\partial}{\partial x}\left[-2\alpha^{3/2} x^2 e^{-\alpha x^2/2} + 2\alpha^{1/2} e^{-\alpha x^2/2}\right] = -4\alpha^{3/2} x e^{-\alpha x^2/2} - 2\alpha^{3/2} x^2 e^{-\alpha x^2/2}(-\alpha x) + 2\alpha^{1/2} e^{-\alpha x^2/2}(-\alpha x)$,

which simplifies to $2\alpha^{5/2} x^3 e^{-\alpha x^2/2} - 4\alpha^{3/2} x e^{-\alpha x^2/2} - 2\alpha^{3/2} x e^{-\alpha x^2/2}$. Recalling the original definition of Ψ_1, we can rewrite this as (if we include the N) $\alpha^2 x^2 \Psi_1 - 3\alpha \Psi_1$. Substituting back into the Schrödinger equation:

$-\dfrac{\hbar^2}{2m}\left(\alpha^2 x^2 \Psi_1 - 3\alpha \Psi_1\right) + \dfrac{1}{2} kx^2 \Psi_1$. Now we can substitute for the definitions of α and k:

$-\dfrac{\hbar^2}{2m}\left(\dfrac{2\pi\nu m}{\hbar}\right)^2 x^2 \Psi_1 + \dfrac{\hbar^2}{2m} \dfrac{6\pi\nu m}{\hbar} \Psi_1 + \dfrac{1}{2}(4\pi^2 \nu^2 m) x^2 \Psi_1 = -\pi^2 \nu^2 m x^2 \Psi_1 + 3\hbar\pi\nu \Psi_1 + \pi^2 \nu^2 m x^2 \Psi_1$

The first and last terms cancel, leaving $3\hbar\pi\nu \Psi_1$. Writing in terms of $h/2\pi$:

$\dfrac{3h\pi\nu}{2\pi} \Psi_1$, which simplifies to $\dfrac{3}{2} h\nu \cdot \Psi_1$ Thus, we have verified that we generate an eigenvalue of

$3/2$ $h\nu$ when we substitute Ψ_1 into the Schrödinger equation.

11.15. $N^2 \displaystyle\int_{-\infty}^{+\infty}\left(x e^{-\alpha x^2/2}\right)^* x e^{-\alpha x^2/2} dx = 1 \qquad N^2 \displaystyle\int_{-\infty}^{+\infty} x^2 e^{-\alpha x^2} dx = 1$ Now we apply one of the integrals

in the integral table from the appendix:

$$N^2 \frac{1}{2}\left(\frac{\pi}{\alpha^3}\right)^{1/2} = 1 \qquad N^2 = 2\left(\frac{\alpha^3}{\pi}\right)^{1/2} \qquad N = 2^{1/2}\left(\frac{\alpha^3}{\pi}\right)^{1/4} \qquad \text{Therefore, the complete}$$

wavefunction is $\Psi_1 = 2^{1/2}\left(\frac{\alpha^3}{\pi}\right)^{1/4} x e^{-\alpha x^2/2}$, which can be shown to be algebraically equivalent to

$\left(\frac{\alpha}{\pi}\right)^{1/4}\left(\frac{1}{2}\right)^{1/2}(2\alpha^{1/2}x)e^{-\alpha x^2/2}$, which is the form of the wavefunction obtained when using equation 11.19.

11.17. The classical turning point is the value of x at which the total energy equals the potential energy. We have expressions for both of those energies, so we simply equate them and solve for x:

$$h\nu(n + \tfrac{1}{2}) = \tfrac{1}{2}kx^2 \qquad x^2 = \frac{2h\nu(n+1/2)}{k} = \frac{(2n+1)h\nu}{k} \qquad x = \pm\sqrt{\frac{(2n+1)h\nu}{k}}$$

11.19. $\dfrac{(2.435 \times 10^{24}\ \text{kg})(2.995 \times 10^{22}\ \text{kg})}{2.435 \times 10^{24}\ \text{kg} + 2.995 \times 10^{22}\ \text{kg}} = 2.959 \times 10^{22}\ \text{kg.}$

11.21. Since we know that the frequency of vibration is inversely proportional to the square root of the reduced mass, we can make the following ratio:

$\nu \propto \dfrac{1}{\sqrt{\mu}}$ and $\nu^* \propto \dfrac{1}{\sqrt{\mu^*}}$ leads us to the following ratio: $\dfrac{\nu}{\nu^*} = \dfrac{1/\sqrt{\mu}}{1/\sqrt{\mu^*}} = \sqrt{\dfrac{\mu^*}{\mu}}$. Now we need

only calculate the reduced masses of the O-H and O-D bonds. Since the expression we derived is a ratio, it doesn't matter what ultimate units the reduced mass is in, so let us simply use values in grams: $\mu(OH) = (1)(16)/(1+16) = 0.941$, while $\mu(OD) = (2)(16)/(2+16) = 1.778$. Therefore,

we get $\dfrac{3650\ \text{cm}^{-1}}{\nu^*} = \sqrt{\dfrac{1.778}{0.941}} = 1.374$ Solve for ν^*: $\nu^* = 2656\ \text{cm}^{-1}$.

11.23. For Ψ_3: $N^2 \displaystyle\int_0^{2\pi}\left(e^{i3\phi}\right)^* e^{i3\phi}\,d\phi = 1 \qquad N^2 \displaystyle\int_0^{2\pi}\left(e^{-i3\phi}\right)e^{i3\phi}\,d\phi = 1 \qquad N^2 \displaystyle\int_0^{2\pi} d\phi = 1$

$N^2\left[\phi\right]_0^{2\pi} = 1 \qquad N^2(2\pi - 0) = 1 \qquad N^2 = \dfrac{1}{2\pi} \qquad N = \dfrac{1}{\sqrt{2\pi}}$

For Ψ_{13}: $N^2 \displaystyle\int_0^{2\pi}\left(e^{i13\phi}\right)^* e^{i13\phi}\,d\phi = 1 \qquad N^2 \displaystyle\int_0^{2\pi}\left(e^{-i13\phi}\right)e^{i13\phi}\,d\phi = 1 \qquad N^2 \displaystyle\int_0^{2\pi} d\phi = 1$

$N^2\left[\phi\right]_0^{2\pi} = 1 \qquad N^2(2\pi - 0) = 1 \qquad N^2 = \dfrac{1}{2\pi} \qquad N = \dfrac{1}{\sqrt{2\pi}}$

Thus, both wavefunctions have the same normalization constant.

11.25. (a) $600 \, \text{kg} \cdot \text{m}^2/\text{s} = \dfrac{m(6.626 \times 10^{-34} \, \text{J} \cdot \text{s})}{2\pi}$ $m \approx 5.69 \times 10^{36}$

(b) $E = \dfrac{(5.69 \times 10^{36})^2 (6.626 \times 10^{-34} \, \text{J} \cdot \text{s})^2}{(2\pi)^2 (2)(25 \, \text{kg})(8 \, \text{m})^2} = 113 \, \text{J}$ Classically, the child would have an

energy of $\dfrac{L^2}{2I} = \dfrac{(600 \, \text{kg} \cdot \text{m}^2/\text{s})^2}{2(25 \, \text{kg})(8 \, \text{m})^2} = 113 \, \text{J}$, the same amount – but this shouldn't be surprising,

since we calculated the 'quantum-mechanical' amount of energy using the same values but in a more roundabout fashion.

11.27. (a) $E(m+1) - E(m) = \dfrac{(m+1)^2 \hbar^2}{2I} - \dfrac{m^2 \hbar^2}{2I} = \dfrac{(2m+1)\hbar^2}{2I}$

(b) If $E(1) - E(0) = 20.7 \, \text{cm}^{-1}$, we can use the formula from part a to determine I, then calculate $E(2) - E(1)$ using that value of I.

$20.7 \, \text{cm}^{-1} = \dfrac{(2 \cdot 0 + 1)(6.626 \times 10^{-34} \, \text{J} \cdot \text{s})^2}{(2\pi)^2 (2) I}$ $I = 2.686 \times 10^{-70} \, \text{J}^2 \text{m}^2/\text{cm}^{-1}$ (we are not reducing

the units, since we're simply using this quantity to recalculate an energy difference in cm^{-1} units)

Now, calculating $E(2) - E(1)$: $E(2) - E(1) = \dfrac{(2 \cdot 1 + 1)(6.626 \times 10^{-34} \, \text{J} \cdot \text{s})^2}{(2\pi)^2 (2)(2.686 \times 10^{-70} \, \text{J}^2 \text{s}^2/\text{cm}^{-1})} = 62.1 \, \text{cm}^{-1}$.

This is not a great agreement with the experimental value of $41.4 \, \text{cm}^{-1}$, suggesting that the 2-D rotational model may not be the best model for HCl.

11.29. Refer to Figure 11.12. The easiest coordinate to relate is the z coordinate. Projecting the r vector into the z axis, we find that the r vector is the hypotenuse of a right triangle. Since the

definition of cosine of an angle is "adjacent side over hypotenuse", we have $\cos\theta = \dfrac{z}{r}$, which

rearranges to $z = r \cos\theta$. The other leg of that triangle, by the way, is $r \sin\theta$, which is drawn in Figure 11.12 as the dotted-line projection of the r vector in the xy plane. This dotted-line projection serves as the hypotenuse of another right triangle. This dotted line's projection into the x axis, the x coordinate, can be related to ϕ through the cosine function, which is "adjacent

leg over hypotenuse". Thus, $\cos\phi = \dfrac{x}{r \sin\theta}$. Therefore, we have $x = r \sin\theta \cos\phi$. The other leg

of that triangle can be related to the y coordinate through the sine function, as "opposite leg over

hypotenuse. Therefore, we have $\sin\phi = \dfrac{y}{r \sin\theta}$, or $y = r \sin\theta \sin\phi$.

11.31. Because a potential energy is a multiplicative operator, if it is simply a constant K, we have:

$$-\dfrac{\hbar^2}{2I}\left(\dfrac{\partial^2}{\partial\theta^2} + \cot\theta \dfrac{\partial}{\partial\theta} + \dfrac{1}{\sin^2\theta}\dfrac{\partial^2}{\partial\phi^2}\right)\Psi + K\Psi = E\Psi$$

$$-\dfrac{\hbar^2}{2I}\left(\dfrac{\partial^2}{\partial\theta^2} + \cot\theta \dfrac{\partial}{\partial\theta} + \dfrac{1}{\sin^2\theta}\dfrac{\partial^2}{\partial\phi^2}\right)\Psi = E\Psi - K\Psi = (E-K)\Psi = E_{new}\Psi.$$

So we could in fact use the same form of the Schrödinger equation if the potential energy were some constant value.

11.33. (a) $\hat{L}^2\Psi_{3,-2} = -\hbar^2\left(\dfrac{\partial^2}{\partial\theta^2} + \cot\theta\,\dfrac{\partial}{\partial\theta} + \dfrac{1}{\sin^2\theta}\dfrac{\partial^2}{\partial\phi^2}\right)\left(\dfrac{\sqrt{105}}{4}\sin^2\theta\cos\theta e^{-2i\phi}\right)$

Let us determine each derivative singly, then combine the three terms and simplify. The first term:

$= \dfrac{\partial}{\partial\theta}\Psi_{3,-2} = \left(\dfrac{\sqrt{105}}{4}\cdot 2\sin\theta(\cos\theta)\cos\theta - \dfrac{\sqrt{105}}{4}\sin^3\theta\right)e^{-2i\phi} = \dfrac{\sqrt{105}}{4}\left(2\sin\theta\cos^2\theta - \sin^3\theta\right)e^{-2i\phi}$

$\dfrac{\partial^2}{\partial\theta^2}\Psi_{3,-2} = \dfrac{\sqrt{105}}{4}\left(2\cos^3\theta - 4\sin^2\theta\cos\theta - 3\sin^2\theta\cos\theta\right)e^{-2i\phi}$

The second term:

$\cot\theta\,\dfrac{\partial}{\partial\theta}\Psi_{3,-2} = \cot\theta\,\dfrac{\sqrt{105}}{4}\left(2\sin\theta\cos^2\theta - \sin^3\theta\right)e^{-2i\phi} = \dfrac{\sqrt{105}}{4}\left(2\cos^3\theta - \sin^2\theta\cos\theta\right)e^{-2i\phi}$

The third term:

$\dfrac{1}{\sin^2\theta}\dfrac{\partial^2}{\partial\phi^2}\Psi_{3,-2} = \dfrac{1}{\sin^2\theta}\left(\dfrac{\sqrt{105}}{4}\sin^2\theta\cos\theta e^{-2i\phi}\cdot(-2i)^2\right) = -4\dfrac{\sqrt{105}}{4}\cos\theta e^{-2i\phi}$

Combining these terms:

$-\hbar^2\left(\dfrac{\sqrt{105}}{4}\left(2\cos^3\theta - 4\sin^2\theta\cos\theta - 3\sin^2\theta\cos\theta + 2\cos^3\theta - \sin^2\theta\cos\theta - 4\cos\theta\right)e^{-2i\phi}\right)$

This simplifies to

$-\hbar^2\left(\dfrac{\sqrt{105}}{4}\left(4\cos^3\theta - 8\sin^2\theta\cos\theta - 4\cos\theta\right)e^{-2i\phi}\right)$ Factoring a cosine out of the first two terms:

$-\hbar^2\left(\dfrac{\sqrt{105}}{4}\cos\theta\left(4\cos^2\theta - 8\sin^2\theta - 4\right)e^{-2i\phi}\right)$ If we recall that $\sin^2\theta + \cos^2\theta = 1$, then we can

substitute the 4 inside the parentheses with $4\sin^2\theta + 4\cos^2\theta$ (but don't forget to distribute the minus sign:

$-\hbar^2\left(\dfrac{\sqrt{105}}{4}\cos\theta\left(4\cos^2\theta - 8\sin^2\theta - 4\sin^2\theta - 4\cos^2\theta\right)e^{-2i\phi}\right)$ The $4\cos^2\theta$ terms cancel, and we

can group the $\sin^2\theta$ terms together to get

$-\hbar^2\left(\dfrac{\sqrt{105}}{4}\cos\theta\left(-12\sin^2\theta\right)e^{-2i\phi}\right) = +12\hbar^2\left(\dfrac{\sqrt{105}}{4}\sin^2\theta\cos\theta e^{-2i\phi}\right) = +12\hbar^2\Psi_{3,-2}$

Therefore, this wavefunction is an eigenfunction of the square angular momentum operator with an eigenvalue of $12\hbar^2$ – as expected from the eigenvalue formula.
(b) This determination is more straightforward because there is only one term in the wavefunction that depends on ϕ: the exponential function. We have:

$$-i\hbar\frac{\partial}{\partial\phi}\Psi_{3,-2} = -i\hbar\frac{\partial}{\partial\phi}\left(\frac{\sqrt{105}}{4}\sin^2\theta\cos\theta e^{-2i\phi}\right) = -i\hbar(-2i)\left(\frac{\sqrt{105}}{4}\sin^2\theta\cos\theta e^{-2i\phi}\right) = -2\hbar\Psi_{3,-2}$$

Therefore, this wavefunction is an eigenfunction of the 2-D angular momentum operator with an eigenvalue of $-2\hbar$ – as expected from the eigenvalue formula.

(c) To determine the energy eigenvalue, technically we have to evaluate the entire three-part derivative in equation 11.43. However, if we recognize that $\hat{H} = \dfrac{\hat{L}^2}{2I}$, we can use the answer to part a because we have already evaluated the eigenvalue to the L^2 operator. Thus, we have

$$\hat{H}\Psi_{3,-2} = \frac{1}{2I}\hat{L}^2\Psi_{3,-2} = \frac{1}{2I}\left(12\hbar^2\right) = \frac{12\hbar^2}{2I} \quad\text{– as expected from the eigenvalue formula.}$$

11.35. (a) $E(\ell+1) - E(\ell) = \dfrac{(\ell+1)(\ell+1+1)\hbar^2}{2I} - \dfrac{(\ell)(\ell+1)\hbar^2}{2I} = \dfrac{(2\ell+2)\hbar^2}{2I}$.

(b) Use the formula from part a to determine the value of I, then use this value to determine the energy difference between the new quantum levels. We have:

$$20.7\,\text{cm}^{-1} = \frac{(2\cdot0+2)(6.626\times10^{-34}\,\text{J}\cdot\text{s})^2}{(2\pi)^2(2)I} \qquad I = 5.370\times10^{-70}\,\text{J}^2\cdot\text{s}^2/\text{cm}^{-1}\ \text{(we are not reducing}$$

the units, since we're simply using this quantity to recalculate an energy difference in cm^{-1} units)

Now, calculating $E(2) - E(1)$: $E(2) - E(1) = \dfrac{(2\cdot1+2)(6.626\times10^{-34}\,\text{J}\cdot\text{s})^2}{(2\pi)^2(2)(5.370\times10^{-70}\,\text{J}^2\text{s}^2/\text{cm}^{-1})} = 41.4\,\text{cm}^{-1}$.

This is exactly the same as the experimental value of $41.4\,\text{cm}^{-1}$, suggesting that the 3-D rotational model is a good model for HCl.

11.37. For the four ℓ values in exercise 11.36:
For $\ell = 5$: $L^2 = \ell(\ell+1)h^2/(2\pi)^2 = 5(5+1)(6.626\times10^{-34}\,\text{J}\cdot\text{s})^2/(2\pi)^2 = 3.34\times10^{-67}\,\text{J}^2\cdot\text{s}^2$. Therefore, the angular momentum is the square root of this: $L = 5.78\times10^{-34}\,\text{J}\cdot\text{s}$.
For $\ell = 6$: $L^2 = \ell(\ell+1)h^2/(2\pi)^2 = 6(6+1)(6.626\times10^{-34}\,\text{J}\cdot\text{s})^2/(2\pi)^2 = 4.67\times10^{-67}\,\text{J}^2\cdot\text{s}^2$. Therefore, the angular momentum is the square root of this: $L = 6.83\times10^{-34}\,\text{J}\cdot\text{s}$.
For $\ell = 7$: $L^2 = \ell(\ell+1)h^2/(2\pi)^2 = 7(7+1)(6.626\times10^{-34}\,\text{J}\cdot\text{s})^2/(2\pi)^2 = 6.23\times10^{-67}\,\text{J}^2\cdot\text{s}^2$. Therefore, the angular momentum is the square root of this: $L = 7.89\times10^{-34}\,\text{J}\cdot\text{s}$.
For $\ell = 8$: $L^2 = \ell(\ell+1)h^2/(2\pi)^2 = 8(8+1)(6.626\times10^{-34}\,\text{J}\cdot\text{s})^2/(2\pi)^2 = 8.01\times10^{-67}\,\text{J}^2\cdot\text{s}^2$. Therefore, the angular momentum is the square root of this: $L = 8.95\times10^{-34}\,\text{J}\cdot\text{s}$.

11.39. Physically, the two wavefunctions are the same except they are reflections through the xy plane. One wavefunction has a positive z component, and the other wavefunction has a negative z component.

11.41. $V = -\dfrac{e^2}{4\pi\varepsilon_0 r} = -\dfrac{(1.602\times10^{-19}\,\text{C})^2}{4\pi(8.854\times10^{-12}\,\text{C}^2/\text{J}\cdot\text{m})(0.529\times10^{-10}\,\text{m})} = -4.36\times10^{-18}\,\text{J}$

11.43. Starting with equation 11.56:

$$\left[-\frac{\hbar^2}{2\mu}\left(\frac{1}{r^2}\frac{\partial}{\partial r}\left(r^2\frac{\partial}{\partial r}\right)+\frac{1}{r^2\sin\theta}\frac{\partial}{\partial\theta}\left(\sin\theta\frac{\partial}{\partial\theta}\right)+\frac{1}{r^2\sin^2\theta}\frac{\partial^2}{\partial\phi^2}\right)+\hat{V}\right]\Psi=E\Psi$$

If $V=0$ and r is constant, we lose two things immediately: the potential energy operator and the first derivative term in r (since r is constant, any derivative with respect to r is zero). We have

$$-\frac{\hbar^2}{2\mu}\left(\frac{1}{r^2\sin\theta}\frac{\partial}{\partial\theta}\left(\sin\theta\frac{\partial}{\partial\theta}\right)+\frac{1}{r^2\sin^2\theta}\frac{\partial^2}{\partial\phi^2}\right)\Psi=E\Psi \quad \text{Now, distributing the derivative with}$$

respect to theta through the parentheses:

$$-\frac{\hbar^2}{2\mu}\left(\frac{1}{r^2\sin\theta}\left(\cos\theta\frac{\partial}{\partial\theta}+\sin\theta\frac{\partial^2}{\partial\theta^2}\right)+\frac{1}{r^2\sin^2\theta}\frac{\partial^2}{\partial\phi^2}\right)\Psi=E\Psi \quad \text{If we factor out the } r^2 \text{ in the}$$

denominator and combine it with μ in the denominator, we get the moment of inertia, I:

$$-\frac{\hbar^2}{2I}\left(\frac{1}{\sin\theta}\left(\cos\theta\frac{\partial}{\partial\theta}+\sin\theta\frac{\partial^2}{\partial\theta^2}\right)+\frac{1}{\sin^2\theta}\frac{\partial^2}{\partial\phi^2}\right)\Psi=E\Psi \quad \text{Finally, if we distribute the sin}\theta \text{ term}$$

in the denominator through, we get

$$-\frac{\hbar^2}{2I}\left(\cot\theta\frac{\partial}{\partial\theta}+\frac{\partial^2}{\partial\theta^2}+\frac{1}{\sin^2\theta}\frac{\partial^2}{\partial\phi^2}\right)\Psi=E\Psi, \text{ which is equation 11.46.}$$

11.45. (a) First, we must convert the wavelengths to wavenumber units:

$$\frac{1}{656.5\,\text{nm}}\times\frac{10^7\,\text{nm}}{1\,\text{cm}}=15{,}230\,\text{cm}^{-1} \qquad\qquad \frac{1}{486.3\,\text{nm}}\times\frac{10^7\,\text{nm}}{1\,\text{cm}}=20{,}560\,\text{cm}^{-1}$$

$$\frac{1}{434.2\,\text{nm}}\times\frac{10^7\,\text{nm}}{1\,\text{cm}}=23{,}030\,\text{cm}^{-1} \qquad\qquad \frac{1}{410.3\,\text{nm}}\times\frac{10^7\,\text{nm}}{1\,\text{cm}}=24{,}370\,\text{cm}^{-1}$$

These transitions are from quantum levels 3, 4, 5, and 6, respectively. Therefore, we can calculate R as follows:

$$15{,}230\,\text{cm}^{-1}=R\left(\frac{1}{2^2}-\frac{1}{3^2}\right) \qquad R=109{,}700\,\text{cm}^{-1}$$

$$20{,}560\,\text{cm}^{-1}=R\left(\frac{1}{2^2}-\frac{1}{4^2}\right) \qquad R=109{,}700\,\text{cm}^{-1}$$

$$23{,}030\,\text{cm}^{-1}=R\left(\frac{1}{2^2}-\frac{1}{5^2}\right) \qquad R=109{,}700\,\text{cm}^{-1}$$

$$24{,}370\,\text{cm}^{-1}=R\left(\frac{1}{2^2}-\frac{1}{6^2}\right) \qquad R=109{,}700\,\text{cm}^{-1}$$

Thus, to four significant figures, all four lines give the same value of the Rydberg constant.
(b) Since $Z=2$ for helium, there is an extra Z^2 term in the numerator of the Rydberg constant for He$^+$, so R would equal $2^2\times109{,}700\,\text{cm}^{-1}=438{,}800\,\text{cm}^{-1}$. Therefore, we would expect the lines to appear at:

$$438{,}800\,\text{cm}^{-1}\left(\frac{1}{2^2}-\frac{1}{3^2}\right)=60{,}940\,\text{cm}^{-1} \text{ or } 164\,\text{nm}$$

$$438{,}800 \text{ cm}^{-1}\left(\frac{1}{2^2} - \frac{1}{4^2}\right) = 82{,}280 \text{ cm}^{-1} \text{ or } 122 \text{ nm}$$

$$438{,}800 \text{ cm}^{-1}\left(\frac{1}{2^2} - \frac{1}{5^2}\right) = 92{,}150 \text{ cm}^{-1} \text{ or } 109 \text{ nm}$$

$$438{,}800 \text{ cm}^{-1}\left(\frac{1}{2^2} - \frac{1}{6^2}\right) = 97{,}510 \text{ cm}^{-1} \text{ or } 103 \text{ nm}$$

11.47. Figure 11.17 in the text gives such a diagram.

11.49. The wavefunction $\Psi_{4,4,0}$ does not exist because ℓ cannot be equal to n. Similarly, a $3f$ subshell does not exist because n would equal 3 and ℓ would equal 3 also, and the constraints on the hydrogen atom wavefunctions are that ℓ must be less than n.

11.51. The setup is:

$$P = \iiint \left(\frac{1}{\pi(0.529\text{A})^3}\right)^{1/2} e^{-r/0.529\text{A}} \cdot \left(\frac{1}{\pi(0.529\text{A})^3}\right)^{1/2} e^{-r/0.529\text{A}} \cdot r^2 \sin\theta\, dr\, d\theta\, d\phi$$

This is a triple integral, where the limits on ϕ are 0 to 2π, the limits on θ are 0 to π, and the limits on r are 0 to 1Å. Thus, we have

$$P = \int_0^{2\pi} d\phi \times \int_0^{\pi} \sin\theta\, d\theta \times \left(\frac{1}{\pi(0.529\text{A})^3}\right) \int_0^{0.1\text{A}} r^2 e^{-2r/0.529\text{A}}\, dr \quad \text{As shown in the text (see example 11.24),}$$

the θ and ϕ integrals collectively equal 4π. Thus, we have

$$P = 4\pi \times \left(\frac{1}{\pi(0.529\text{A})^3}\right) \int_0^{0.1\text{A}} r^2 e^{-2r/0.529\text{A}}\, dr \quad \text{Using the integral table in the appendix, we can solve}$$

the integral and evaluate:

$$P = 4\pi \left(\frac{1}{\pi(0.529\text{A})^3}\right) \times \left[e^{-2r/0.529\text{A}}\left(\frac{r^2}{(-2/0.529\text{A})} - \frac{2r}{(-2/0.529\text{A})^2} + \frac{2}{(-2/0.529\text{A})^3}\right)\right]_0^{0.1\text{A}}$$

$$P = 4\pi\left(2.150221\text{A}^{-3}\right) \times \left[e^{-2(0.1)/0.529}\left(-\frac{(0.1\text{A})^2(0.529\text{A})}{2} - \frac{2(0.1\text{A})(0.529\text{A})^2}{4} - \frac{2(0.529\text{A})^3}{8}\right)\right.$$

$$\left. - e^0\left(0 - 0 - \frac{2(0.529\text{A})^3}{8}\right)\right]$$

$$P = 4\pi(2.150221\text{A}^{-3}) \times [0.68518(-0.002645\text{A}^3 - 0.01399\text{A}^3 - 0.03701\text{A}^3)] + 0.03701\text{A}^3$$
$$P = 0.0068, \text{ or } 0.68\%.$$

11.53. (a) For Ψ_{2s}, there will be one radial node and no angular nodes, for a total of one node. (b) For Ψ_{3s}, there will be two radial nodes and no angular nodes, for a total of two nodes. (c) For Ψ_{3p}, there will be one radial node and one angular node, for a total of two nodes. (d) For Ψ_{4f}, there will be no radial nodes and three angular nodes, for a total of three nodes. (e) For Ψ_{6g}, there will be no radial nodes and five angular nodes, for a total of five nodes. (f) For Ψ_{7s}, there will be six radial nodes and no angular nodes, for a total of six nodes.

11.55. $a = \dfrac{4\pi\varepsilon_0\hbar^2}{\mu e^2} = \dfrac{4\pi(8.854\times10^{-12}\ \text{C}^2/\text{J}\cdot\text{m})(6.626\times10^{-34}\ \text{J}\cdot\text{s})^2}{(2\pi)^2(9.104\times10^{-31}\ \text{kg})(1.602\times10^{-19}\ \text{C})} = 5.29585\times10^{-11}\ \text{m}$

11.57. According to equation 11.67, $\Psi_{2p_x} = \dfrac{1}{\sqrt{2}}\left(\Psi_{2p_{+1}} + \Psi_{2p_{-1}}\right)$. Using the forms of the wavefunctions from Table 11.4, we have

$$\Psi_{2p_x} = \frac{1}{\sqrt{2}}\left(\frac{1}{8}\left(\frac{2Z^3}{\pi a^3}\right)^{1/2}\frac{Zr}{a}e^{-Zr/2a}\sin\theta e^{+i\phi} + \frac{1}{8}\left(\frac{2Z^3}{\pi a^3}\right)^{1/2}\frac{Zr}{a}e^{-Zr/2a}\sin\theta e^{-i\phi}\right)$$

$$= \frac{1}{\sqrt{2}}\frac{1}{8}\left(\frac{2Z^3}{\pi a^3}\right)^{1/2}\frac{Zr}{a}e^{-Zr/2a}\sin\theta\left(e^{+i\phi} + e^{-i\phi}\right)$$ Now we use the fact that $\cos\phi = \dfrac{e^{i\phi} + e^{-i\phi}}{2}$ and

substitute: $\Psi_{2p_x} = \dfrac{1}{4\sqrt{2}}\left(\dfrac{2Z^3}{\pi a^3}\right)^{1/2}\dfrac{Zr}{a}e^{-Zr/2a}\sin\theta\cos\phi$.

Also according to equation 11.67, $\Psi_{2p_y} = -\dfrac{i}{\sqrt{2}}\left(\Psi_{2p_{+1}} - \Psi_{2p_{-1}}\right)$. Using the forms of the wavefunctions from Table 11.4, we have

$$\Psi_{2p_x} = -\frac{i}{\sqrt{2}}\left(\frac{1}{8}\left(\frac{2Z^3}{\pi a^3}\right)^{1/2}\frac{Zr}{a}e^{-Zr/2a}\sin\theta e^{+i\phi} - \frac{1}{8}\left(\frac{2Z^3}{\pi a^3}\right)^{1/2}\frac{Zr}{a}e^{-Zr/2a}\sin\theta e^{-i\phi}\right)$$

$$= -\frac{i}{\sqrt{2}}\frac{1}{8}\left(\frac{2Z^3}{\pi a^3}\right)^{1/2}\frac{Zr}{a}e^{-Zr/2a}\sin\theta\left(e^{+i\phi} - e^{-i\phi}\right)$$ Now we use the fact that $\sin\phi = \dfrac{e^{i\phi} - e^{-i\phi}}{2i}$ and

substitute: $\Psi_{2p_y} = \dfrac{1}{4\sqrt{2}}\left(\dfrac{2Z^3}{\pi a^3}\right)^{1/2}\dfrac{Zr}{a}e^{-Zr/2a}\sin\theta\sin\phi$. Note how the two i terms and the negative sign canceled.

11.59. The combinations are:

$\Psi_{3d_1} = \dfrac{1}{\sqrt{2}}\left(\Psi_{3d_{+1}} + \Psi_{3d_{-1}}\right)$ $\qquad\qquad$ $\Psi_{3d_2} = -\dfrac{i}{\sqrt{2}}\left(\Psi_{3d_{+1}} - \Psi_{3d_{-1}}\right)$

$\Psi_{3d_3} = \dfrac{1}{\sqrt{2}}\left(\Psi_{3d_{+2}} + \Psi_{3d_{-2}}\right)$ $\qquad\qquad$ $\Psi_{3d_4} = -\dfrac{i}{\sqrt{2}}\left(\Psi_{3d_{+2}} - \Psi_{3d_{-2}}\right)$

$\Psi_{3d_5} = \Psi_{3d_0}$

The numerical labels on these composite wavefunctions are arbitrary; they are actually given Cartesian coordinate labels that reflect their distribution in three-dimensional space. Note that one of the wavefunctions is a pure eigenfunction, rather than a combination of two eigenfunctions.

CHAPTER 12. ATOMS AND MOLECULES

12.1. Silver atoms have a single unpaired electron in their valence shell. Thus, there is an unbalanced overall spin to the atoms, making them susceptible to magnetic fields.

12.3. The total mass being converted to energy is the mass of two electrons, or $2\times9.109\times10^{-31}$ kg $= 1.822\times10^{-30}$ kg. Using Einstein's equation:
$E = mc^2 = (1.822\times10^{-30}$ kg$)(2.9979\times10^8$ m/s$)^2 = 1.638\times10^{-13}$ J $\times 6.02\times10^{23} = 9.86\times10^{10}$ J/mol

12.5. (a) The quantum number s represents the total spin angular momentum quantum number. For all electrons, s is ½ (and is a characteristic of the particle). The quantum number m_s represents the z component of the total spin angular momentum. For electrons, it can be either $+\frac{1}{2}$ or $-\frac{1}{2}$. (b) If a particle has $s = 0$, then m_s can only equal 0. If a particle has $s = 2$, then m_s can be -2, -1, 0, 1, or 2. If a particle has $s = 3/2$, then m_s can equal $-3/2$, $-1/2$, $1/2$, or $3/2$.

12.7. $\hat{H} = -\dfrac{\hbar^2}{2\mu}\left(\nabla^2_{e\#1} + \nabla^2_{e\#2} + \nabla^2_{e\#3}\right) - \dfrac{3e^2}{4\pi\varepsilon_0 r_1} - \dfrac{3e^2}{4\pi\varepsilon_0 r_2} - \dfrac{3e^2}{4\pi\varepsilon_0 r_3} + \dfrac{e^2}{4\pi\varepsilon_0 r_{12}} + \dfrac{e^2}{4\pi\varepsilon_0 r_{13}} + \dfrac{e^2}{4\pi\varepsilon_0 r_{23}}$

The last three terms are what make this Hamiltonian non-separable.

12.9. Acceptable wavefunctions are linear combinations (i.e. sums and/or differences) of individual spatial wavefunctions because this way the eigenvalue energy is the correct value. If the wavefunction were constructed as the product of spatial wavefunctions, then the resulting energy eigenvalue would be the product of the energies of the individual wavefunctions, which is not the correct energy eigenvalue for the atom.

12.11. Li^+, with only two electrons, would have the same wavefunction as the helium atom.

Therefore we simply use that wavefunction: $\Psi(1,2) = \dfrac{1}{\sqrt{2}}\left[(1s_1\alpha)(1s_2\beta) - (1s_1\beta)(1s_2\alpha)\right]$.

12.13. (a) $\Psi(Be) = \dfrac{1}{\sqrt{24}}\begin{vmatrix} 1s_1\alpha & 1s_1\beta & 2s_1\alpha & 2s_1\beta \\ 1s_2\alpha & 1s_2\beta & 2s_2\alpha & 2s_2\beta \\ 1s_3\alpha & 1s_3\beta & 2s_3\alpha & 2s_3\beta \\ 1s_4\alpha & 1s_4\beta & 2s_4\alpha & 2s_4\beta \end{vmatrix}$

$\Psi(B) = \dfrac{1}{\sqrt{120}}\begin{vmatrix} 1s_1\alpha & 1s_1\beta & 2s_1\alpha & 2s_1\beta & 2p_{x,1}\alpha \\ 1s_2\alpha & 1s_2\beta & 2s_2\alpha & 2s_2\beta & 2p_{x,2}\alpha \\ 1s_3\alpha & 1s_3\beta & 2s_3\alpha & 2s_3\beta & 2p_{x,3}\alpha \\ 1s_4\alpha & 1s_4\beta & 2s_4\alpha & 2s_4\beta & 2p_{x,4}\alpha \\ 1s_5\alpha & 1s_5\beta & 2s_5\alpha & 2s_5\beta & 2p_{x,5}\alpha \end{vmatrix}$. Keep in mind that, because the six p orbitals

are degenerate, the last column could have been labeled $2p_x\beta$, $2p_y\alpha$, $2p_y\beta$, $2p_z\alpha$, or $2p_z\beta$.
(b) There are six possible determinants for the C atom that have the two unpaired p electrons having the same spin. There are also six possible determinants for the F atom, depending on which spin-orbital is not used (analogous to the wavefunction for Be in part a above).

12.15. Atoms that don't follow the aufbau principle exactly include Cr, Cu, Nb, Mo, Ru, Rh, Pb, Ag, La, Ce, Gc, Pt, Au,, Ac, Th, Pa, U, Np, and Cm. Note that none of these atoms are main group elements; they are transition metals or inner transition (i.e. lanthanide or actinide) elements.

12.17. (a) Li will have 6 possibilities for its unfilled valence shell: $2p_x{}^1\alpha$, $2p_x{}^1\beta$, $2p_y{}^1\alpha$, $2p_y{}^1\beta$, $2p_z{}^1\alpha$, and $2p_z{}^1\beta$. (b) C will have six possibilities for its unfilled valence shell: $2p_x{}^1\alpha\ 2p_y{}^1\alpha$, $2p_x{}^1\alpha\ 2p_z{}^1\alpha$, $2p_y{}^1\alpha\ 2p_z{}^1\alpha$, $2p_x{}^1\beta\ 2p_y{}^1\beta$, $2p_x{}^1\beta\ 2p_z{}^1\beta$, and $2p_y{}^1\beta\ 2p_z{}^1\beta$. (c) K will have six possibilities for its unfilled valence shell: $4p_x{}^1\alpha$, $4p_x{}^1\beta$, $4p_y{}^1\alpha$, $4p_y{}^1\beta$, $4p_z{}^1\alpha$, and $4p_z{}^1\beta$. (d) Be will only have one possibility for its excited state, since both electrons have been excited to an s orbital: $3s_1\alpha\ 3s_2\beta$. (e) U will have 6 possibilities for its unfilled valence shell: $7p_x{}^1\alpha$, $7p_x{}^1\beta$, $7p_y{}^1\alpha$, $7p_y{}^1\beta$, $7p_z{}^1\alpha$, and $7p_z{}^1\beta$. Note the similarity in the answers for Li, K, and U.

12.19. The ground-state harmonic oscillator wavefunction is found in Chapter 11.

$$\int_{-\infty}^{+\infty}\left[\left(\frac{\alpha}{\pi}\right)^{1/4}e^{-\alpha x^2/2}\right]^* \cdot cx^4 \cdot \left(\frac{\alpha}{\pi}\right)^{1/4}e^{-\alpha x^2/2}dx = c\left(\frac{\alpha}{\pi}\right)^{1/2}\int_{-\infty}^{+\infty}x^4 e^{-\alpha x^2}dx.$$

According to the table in the Appendix, there is an appropriate form to use if the integral limits are 0 to ∞. Since the function in the integral is even, we can split the interval and multiply the value of the integral by 2. We get:

$$= 2\cdot c\left(\frac{\alpha}{\pi}\right)^{1/2}\left(\frac{1\cdot 3}{2^3\cdot \alpha^2}\right)\left(\frac{\pi}{\alpha}\right)^{1/2}\quad \text{The two square root terms cancel. Simplifying:}$$

$$= \frac{3c}{4\alpha^2}, \text{ where } c \text{ is the anharmonicity constant.}$$

12.21. To determine a_3, follow the same procedure used in Example 12.10 but use Ψ_1 and Ψ_3 for the particle-in-a-box:

$$a_3 = \frac{\int_0^a \Psi_3{}^* \cdot kx \cdot \Psi_1 dx}{E_1 - E_3} = \frac{\int_0^a \sqrt{\frac{2}{a}}\sin\frac{3\pi x}{a}\cdot kx \cdot \sqrt{\frac{2}{a}}\sin\frac{1\pi x}{a}dx}{\left(\frac{1^2 h^2}{8ma^2} - \frac{3^2 h^2}{8ma^2}\right)} = \frac{\frac{2k}{a}\int_0^a x\sin\frac{3\pi x}{a}\sin\frac{1\pi x}{a}dx}{\left(-\frac{h^2}{ma^2}\right)}$$

Again, we use the trigonometric identity $\sin ax\cdot\sin bx = \frac{1}{2}[\cos(a-b)x - \cos(a+b)x]$ and substitute:

$$a_3 = -\frac{2kma}{h^2}\cdot\frac{1}{2}\int_0^a\left(x\cos\frac{2\pi x}{a} - x\cos\frac{4\pi x}{a}\right)dx\quad \text{This expression can be integrated:}$$

$$= -\frac{kma}{h^2}\left[\frac{a^2}{\pi^2}\cos\frac{2\pi x}{a} + \frac{ax}{\pi}\sin\frac{2\pi x}{a} - \frac{a^2}{16\pi^2}\cos\frac{4\pi x}{a} - \frac{ax}{4\pi}\sin\frac{4\pi x}{a}\right]_0^a\quad \text{Evaluating at the limits:}$$

$$a_3 = -\frac{kma}{h^2}\left[\frac{a^2}{\pi^2}(1) + \frac{a^2}{\pi}(0) - \frac{a^2}{16\pi^2}(1) - \frac{0}{4\pi}(0) - [0 + 0 - 0 - 0]\right] = -\frac{15kma^3}{16\pi^2 h^2}.$$

From the Example, we found that $k = 1\times10^{-7}$ kg·m/s^2, $a = 1.15\times10^{-10}$ m, so we can evaluate this expression explicitly for an electron: $a_3 = -0.02998$.

12.23. The integral we need to evaluate is

$$\int_{r,\theta,\phi} \frac{1}{8}\left(\frac{2}{\pi a^3}\right)^{1/2} \frac{r}{a} e^{-r/2a} \cos\theta \cdot eEr\cos\theta \cdot \left(\frac{1}{\pi a^3}\right)^{1/2} e^{-r/a} \cdot r^2 \sin\theta \, dr \, d\theta \, d\phi \quad \text{(Note that we have the}$$

$2p_z$ orbital as one wavefunction and the $1s$ orbital as the other wavefunction.) Everything is multiplicative, so we can separate this integral into three integrals as follows:

$$= \frac{\sqrt{2}eE}{8\pi a^4} \int_0^{2\pi} d\phi \cdot \int_0^{\pi} \cos^2\theta \sin\theta \, d\theta \cdot \int_0^{\infty} r^4 e^{-3r/2a} dr \ . \quad \text{The integral over } \phi \text{ is simply } 2\pi. \text{ The integral}$$

over θ is easy to evaluate as a power function of cosine:

$$\int_0^{\pi} \cos^2\theta \sin\theta \, d\theta = -\frac{1}{3}\cos^3\theta\Big|_0^{\pi} = -\frac{1}{3}(-1-1) = \frac{2}{3}$$

The integral involving r can also be evaluated using a formula from the appendix:

$$\int_0^{\infty} r^4 e^{-3r/2a} dr = \frac{4!}{(3/2a)^5} = \frac{24 \cdot 32a^5}{243} = \frac{256a^5}{81}\ .$$

Combining the three integral solutions with the collection of constants, we get that

$$\langle E_1 \rangle = \frac{\sqrt{2}eE}{8\pi a^4} \cdot 2\pi \cdot \frac{2}{3} \cdot \frac{256a^5}{81} = \frac{128\sqrt{2}eaE}{243}\ .$$

12.25. Starting with $\dfrac{\int (c_{a,1}\Psi_1 + c_{a,2}\Psi_2)^* \hat{H}(c_{a,1}\Psi_1 + c_{a,2}\Psi_2)d\tau}{\int (c_{a,1}\Psi_1 + c_{a,2}\Psi_2)^* (c_{a,1}\Psi_1 + c_{a,2}\Psi_2)d\tau}$, we distribute the operator and

multiply through to get the individual integrals:

$$= \frac{\int (c_{a,1}\Psi_1 + c_{a,2}\Psi_2)^* (\hat{H}c_{a,1}\Psi_1 + \hat{H}c_{a,2}\Psi_2)d\tau}{\int (c_{a,1}\Psi_1 + c_{a,2}\Psi_2)^* (c_{a,1}\Psi_1 + c_{a,2}\Psi_2)d\tau}$$

$$= \frac{\int (c_{a,1}\Psi_1)^* \hat{H}c_{a,1}\Psi_1 + (c_{a,2}\Psi_2)^* \hat{H}c_{a,1}\Psi_1 + (c_{a,1}\Psi_1)^* \hat{H}c_{a,2}\Psi_2 + (c_{a,2}\Psi_2)^* \hat{H}c_{a,2}\Psi_2 d\tau}{\int (c_{a,1}\Psi_1)^* c_{a,1}\Psi_1 + (c_{a,2}\Psi_2)^* c_{a,1}\Psi_1 + (c_{a,1}\Psi_1)^* c_{a,2}\Psi_2 + (c_{a,2}\Psi_2)^* c_{a,2}\Psi_2 d\tau}$$

The integral sign can be distributed through to all of the terms:

$$= \frac{\int (c_{a,1}\Psi_1)^* \hat{H}c_{a,1}\Psi_1 d\tau + \int (c_{a,2}\Psi_2)^* \hat{H}c_{a,1}\Psi_1 d\tau + \int (c_{a,1}\Psi_1)^* \hat{H}c_{a,2}\Psi_2 d\tau + \int (c_{a,2}\Psi_2)^* \hat{H}c_{a,2}\Psi_2 d\tau}{\int (c_{a,1}\Psi_1)^* c_{a,1}\Psi_1 d\tau + \int (c_{a,2}\Psi_2)^* c_{a,1}\Psi_1 d\tau + \int (c_{a,1}\Psi_1)^* c_{a,2}\Psi_2 d\tau + \int (c_{a,2}\Psi_2)^* c_{a,2}\Psi_2 d\tau}$$

Now the constants can be brought outside the integral signs:

$$= \frac{c_{a,1}^2 \int \Psi_1^* \hat{H}\Psi_1 d\tau + c_{a,1}c_{a,2} \int \Psi_2^* \hat{H}\Psi_1 d\tau + c_{a,1}c_{a,2} \int \Psi_1^* \hat{H}\Psi_2 d\tau + c_{a,2}^2 \int \Psi_2^* \hat{H}\Psi_2 d\tau}{c_{a,1}^2 \int \Psi_1^* \Psi_1 d\tau + c_{a,1}c_{a,2} \int \Psi_2^* \Psi_1 d\tau + c_{a,1}c_{a,2} \int \Psi_1^* \Psi_2 d\tau + c_{a,2}^2 \int \Psi_2^* \Psi_2 d\tau}$$

At this point the definitions from equation 12.28 can be applied, and the expression becomes

$$= \frac{c_{a,1}^2 H_{11} + c_{a,1}c_{a,2}H_{12} + c_{a,1}c_{a,2}H_{12} + c_{a,2}^2 H_{22}}{c_{a,1}^2 S_{11} + c_{a,1}c_{a,2}S_{12} + c_{a,1}c_{a,2}S_{12} + c_{a,2}^2 S_{22}} = \frac{c_{a,1}^2 H_{11} + 2c_{a,1}c_{a,2}H_{12} + c_{a,2}^2 H_{22}}{c_{a,1}^2 S_{11} + 2c_{a,1}c_{a,2}S_{12} + c_{a,2}^2 S_{22}}$$

This is equation 12.29.

12.27. An "effective nuclear charge" approach to determine appropriate wavefunctions for H atoms is unnecessary for two reasons. First, since we can solve the Schrödinger equation analytically for the H atom, we have no need of approximation methods. Second – and perhaps more to the point regarding Example 12.11 – hydrogen atoms only have one electron, so there are no additional electrons acting to "shield" the nucleus from other electrons.

12.29. (a) Using equation 12.31, the non-trivial solution is found by evaluating the determinant $\begin{vmatrix} -15 - E \cdot 1 & -2.5 - E \cdot 0 \\ -2.5 - E \cdot 0 & -4 - E \cdot 1 \end{vmatrix} = 0$. Because the basis wavefunctions are orthonormal, we know that $S_{11} = S_{22} = 1$ and $S_{12} = S_{21} = 0$. We can simplify the determinant to $\begin{vmatrix} -15 - E & -2.5 \\ -2.5 & -4 - E \end{vmatrix} = 0$. We expand the determinant to get the quadratic equation $E^2 + 19E + 53.75 = 0$. Using the quadratic formula, we can determine the two roots for E:

$$E = \frac{-19 \pm \sqrt{391 - 4(1)(53.75)}}{2} \qquad E = -3.46 \text{ or } -15.54$$

(b) The energies of the real system are still rather close to the ideal energies, but are farther away than in Example 12.12. The reason is the larger values of H_{12} and H_{21}. As those values get larger, the energies of the real system deviate more and more from those of the ideal system.

12.31. Given that any trial wavefunction ϕ can be written as a linear combination of ideal wavefunctions Ψ_i: $\phi = \sum c_i \Psi_i$, where $\hat{H}\Psi_i = E_i \Psi_i$. We need to show that $\int \phi^* \hat{H} \phi \, d\tau \geq E_1$, where E_1 is the ground-state energy of the system. We start by substituting the linear combination into the integral: $\int \left(\sum c_i \Psi_i \right)^* \hat{H} \left(\sum c_j \Psi_j \right) d\tau \geq E_1$. We can distribute the Hamiltonian operator through the second summation: $\int \left(\sum c_i \Psi_i \right)^* \left(\sum c_j \hat{H} \Psi_j \right) d\tau \geq E_1$. Since the Ψ_j's are eigenfunctions of the Hamiltonian operator, we can substitute $E_j \Psi_j$ in the second summation: $\int \left(\sum c_i \Psi_i \right)^* \left(\sum c_j E_j \Psi_j \right) d\tau \geq E_1$. Now we rearrange the summation and integral signs, bringing the constants outside of the integral: $\sum_i \sum_j c_i^* c_j E_j \int \Psi_i^* \Psi_j \, d\tau \geq E_1$. Since the ideal wavefunctions are orthogonal, the integral is exactly zero except for the terms where $i = j$. This has the ultimate effect of eliminating one of the sums and making all of the subscript labels the same. Thus, we get, ultimately, that $\int \phi^* \hat{H} \phi \, d\tau = \sum_j c_j^* c_j E_j \geq E_1$ as the variation theorem we are trying to prove.

The coefficients themselves may be positive or negative, but in either case, multiplying the coefficient by itself will always yield a positive value. That is, $c_j^* c_j$ is always greater than zero. Since we know that $E_j \geq E_1$ (that is, any arbitrary energy value is either the ground-state energy or higher than the ground state energy), we can multiply E_j and E_1 by $c_j^* c_j$, a positive number, and still have the same sign on the relationship: $c_j^* c_j Ej \geq c_j^* c_j E_1$. Furthermore, if we sum over all j's, the summation on the left will always be greater than or equal to the summation on the right because the left sum is the sum of larger individual terms:

$\sum_j c_j^* c_j E_j \geq \sum_j c_j^* c_j E_1$, or $\sum_j c_j^* c_j E_j \geq E_1 \sum_j c_j^* c_j$. Thus, we can substitute this last expression into our last integral equation from the last paragraph to get

$$\int \phi^* \hat{H} \phi \, d\tau = \sum_j c_j^* c_j E_j \geq E_1 \sum_j c_j^* c_j .$$

Finally, if the trial wavefunction is normalized (which it should be), we have the following relationships: $1 = \int \phi^* \phi \, d\tau = \int \left(\sum_i c_i \Psi_i \right)^* \left(\sum_j c_j \Psi_j \right) d\tau$, which can be rearranged as $1 = \sum_i \sum_j c_i^* c_j \int \Psi_i^* \Psi_j \, d\tau$. Again, the original Ψ wavefunctions are orthonormal, which means that every integral is zero except for when $i = j$, in which case the integral equals 1. This again has the effect of eliminating one of the summations, and we are left with $1 = \sum_j c_j^* c_j$.

Substituting this into the right side of the last inequality in the previous paragraph yields $\int \phi^* \hat{H} \phi \, d\tau \geq E_1$, which is the variation theorem: any trial wavefunction will have an average energy equal to or greater than then true ground-state energy of the system.

12.33. Mathematically, the Born-Oppenheimer approximation is given by equation 12.36: $\Psi_{molecule} \approx \Psi_{nuc} \cdot \Psi_{el}$. In words, the Born-Oppenheimer approximation is the assumption that electronic motion can be modeled as if the nuclei were not moving, because electrons more so much faster than the much-heavier nuclei.

12.35. Using the expressions in equation 12.45:
$\Delta E = E_2 - E_1 = \dfrac{H_{22} - H_{12}}{1 - S_{12}} - \dfrac{H_{11} + H_{12}}{1 + S_{12}}$. By combining the denominators, cross-multiplying and collecting terms, you can easily show that this expression is equivalent to
$\Delta E = \dfrac{H_{22} - 2H_{12} - H_{11} + (H_{22} + H_{11})S_{12}}{1 - S_{12}^2}$. For H_2^+, $H_{11} = H_{22}$, so we can simplify further:
$\Delta E = \dfrac{2H_{11}S_{12} - 2H_{12}}{1 - S_{12}^2}$

12.37. The bond order for the first excited state of H_2^+ is $-1/2$. Since this state is known to be a dissociative state (that is, in this electronic state the two nuclei move apart spontaneously to achieve a lower energy), we can deduce that any diatomic molecule that has a negative bond order is inherently unstable, spontaneously increasing its internuclear distance to achieve a lower (and more stable) energy.

12.39. Very simply, the first wavefunction in equation 12.43 is the bonding orbital because its corresponding energy is lower than the second wavefunction in equation 12.43.

12.41. O_2^{2+}: $(\sigma 1s)^2 (\sigma^* 1s)^2 (\sigma 2s)^2 (\sigma^* 2s)^2 (\sigma 2p_z)^2 (\pi 2p_{x,y})^4$
O_2^-: $(\sigma 1s)^2 (\sigma^* 1s)^2 (\sigma 2s)^2 (\sigma^* 2s)^2 (\sigma 2p_z)^2 (\pi 2p_{x,y})^4 (\pi^* 2p_{x,y})^3$
O_2^{2-}: $(\sigma 1s)^2 (\sigma^* 1s)^2 (\sigma 2s)^2 (\sigma^* 2s)^2 (\sigma 2p_z)^2 (\pi 2p_{x,y})^4 (\pi^* 2p_{x,y})^4$

12.43. According to Figure 12.24, NO has a bond order of 5/2.

CHAPTER 13. INTRODUCTION TO SYMMETRY IN QUANTUM MECHANICS

13.1. The more symmetry elements an object has, the simpler it is to mathematically describe the three-dimensional shape of an object. Thus, the phrase "higher symmetry" ultimately translates into mathematical simplicity.

13.3. Consult Figure 13.14 for the flow chart on determining the symmetry of an object. (a) A blank sheet of paper would have D_{2h} symmetry. (b) A three-hole sheet of paper would have C_{2v} symmetry. (c) A baseball (with stitching) should have C_{2v} symmetry. (d) A round pencil would have $C_{\infty v}$ symmetry. (e) The Eiffel Tower would have C_{4v} symmetry. (f) A book (without printing) would have C_{2v} symmetry. (g) A human body (without considering the gross anatomical features) would have C_s symmetry, since the front and back are different. (h) A starfish would have D_{5h} symmetry. (i) an unpainted stop sign would have D_{8h} symmetry.

13.5. (a) Yes, these two elements can constitute a complete group. (b). Yes, these two elements can constitute a complete group. (c) These elements are not a complete group, since the identity element E is missing. (d) These elements are not a complete group since the inverse of C_3, C_{-3} or $C_3{}^2$, is missing.

13.7. (a) Since S_n involves C_n followed by reflection in a plane perpendicular to the axis, the matrix would be the same as C_n but with a -1 in the lower right corner:

$$S_n = \begin{bmatrix} \cos\theta & \sin\theta & 0 \\ -\sin\theta & \cos\theta & 0 \\ 0 & 0 & -1 \end{bmatrix}.$$ (b) Inversion involves changing the sign on every coordinate.

Therefore, $i = \begin{bmatrix} -1 & 0 & 0 \\ 0 & -1 & 0 \\ 0 & 0 & -1 \end{bmatrix}.$

13.9. (a) The following shows a multiplication table for the C_{3v} point group (which you can verify by using a molecule like NH_3 as an example):

	E	C_3	$C_3{}^2$	σ_v	σ_v'	σ_v''
E	E	C_3	$C_3{}^2$	σ_v	σ_v'	σ_v''
C_3	C_3	$C_3{}^2$	E	σ_v'	σ_v''	σ_v
$C_3{}^2$	$C_3{}^2$	E	C_3	σ_v''	σ_v	σ_v'
σ_v	σ_v	σ_v''	σ_v'	E	$C_3{}^2$	C_3
σ_v'	σ_v'	σ_v	σ_v''	C_3	E	$C_3{}^2$
σ_v''	σ_v''	σ_v'	σ_v	$C_3{}^2$	C_3	E

Since only the original symmetry elements appear in the multiplication table, we can state that the closure property is satisfied.
(b) Using the above table to substitute sequentially: $\sigma_v(EC_3) = \sigma_v C_3 = \sigma_v'$.
Also, $(\sigma_v E)C_3 = \sigma_v C_3 = \sigma_v'$. Thus, the associative law is satisfied.

13.11. (a) In C_{2v}, $C_2\sigma_v = \sigma_v'$. (b) In C_{2h}, $iC_2 = \sigma_h$. (c) In D_{6h}, $C_6\sigma_h = S_6$. (d) In D_{2d}, $C_2C_2' = S_4$. (e) In O_h, $iS_4 = C_4^3$.

13.13. Porphine has D_{2h} symmetry as a neutral molecule. However, if a metal ion is substituted into the center of the porphine ring, with the simultaneous loss of the two H atoms on the ring nitrogens, the symmetry becomes D_{4h}.

13.15. The tetrahedron has T_d symmetry: E, $8C_3$, $3C_2$, $6S_4$, and $6\sigma_d$. The cube and the octahedron have O_h symmetry: E, $8C_3$, $3C_2$, $6C_4$, $6C_2'$, i, $8S_6$, $3\sigma_h$, $6S_4$, and $6\sigma_d$.

13.17. Using the scheme in Figure 13.14: (a) H_2O_2 has C_2 symmetry. (b) Allene has D_{2d} symmetry. (c) d-Glycine has C_1 symmetry. (d) ℓ-Glycine also has C_1 symmetry. (e) cis-Dichloroethylene has C_{2v} symmetry. (f) trans-Dichloroethylene has C_{2h} symmetry. (g) Toluene has C_1 symmetry. (h) 1,3-Cyclohexadiene has C_{2v} symmetry.

13.19. (a) The wavefunctions of deuterium oxide have C_{2v} symmetry. (b) The wavefunctions of boron trichloride have D_{3h} symmetry. (c) The wavefunctions of methylene chloride have C_{2v} symmetry.

13.21. (a) The possible formulas would be C_4H_4, C_8H_8, and $C_{20}H_{20}$. There wouldn't be a hydrocarbon equivalent to the octahedron or icosahedron. (b) You should be able to verify that C_4H_4 has all the symmetry elements of the T_d point group, while C_8H_8 has all of the symmetry elements of the O_h point group.

13.23. Of the listed species, the following will not have a permanent dipole moment: phosphorus pentachloride, boron trifluoride, diborane, methane, carbon tetrachloride, 2,2-dimethylpropane, and cubane.

$$
13.25. \quad E \begin{bmatrix} x_N \\ y_N \\ z_N \\ x_{H1} \\ y_{H1} \\ z_{H1} \\ x_{H2} \\ y_{H2} \\ z_{H2} \\ x_{H3} \\ y_{H3} \\ z_{H3} \end{bmatrix} = \begin{bmatrix} 1 & 0 & 0 & 0 & 0 & 0 & 0 & 0 & 0 & 0 & 0 & 0 \\ 0 & 1 & 0 & 0 & 0 & 0 & 0 & 0 & 0 & 0 & 0 & 0 \\ 0 & 0 & 1 & 0 & 0 & 0 & 0 & 0 & 0 & 0 & 0 & 0 \\ 0 & 0 & 0 & 1 & 0 & 0 & 0 & 0 & 0 & 0 & 0 & 0 \\ 0 & 0 & 0 & 0 & 1 & 0 & 0 & 0 & 0 & 0 & 0 & 0 \\ 0 & 0 & 0 & 0 & 0 & 1 & 0 & 0 & 0 & 0 & 0 & 0 \\ 0 & 0 & 0 & 0 & 0 & 0 & 1 & 0 & 0 & 0 & 0 & 0 \\ 0 & 0 & 0 & 0 & 0 & 0 & 0 & 1 & 0 & 0 & 0 & 0 \\ 0 & 0 & 0 & 0 & 0 & 0 & 0 & 0 & 1 & 0 & 0 & 0 \\ 0 & 0 & 0 & 0 & 0 & 0 & 0 & 0 & 0 & 1 & 0 & 0 \\ 0 & 0 & 0 & 0 & 0 & 0 & 0 & 0 & 0 & 0 & 1 & 0 \\ 0 & 0 & 0 & 0 & 0 & 0 & 0 & 0 & 0 & 0 & 0 & 1 \end{bmatrix} \begin{bmatrix} x_N \\ y_N \\ z_N \\ x_{H1} \\ y_{H1} \\ z_{H1} \\ x_{H2} \\ y_{H2} \\ z_{H2} \\ x_{H3} \\ y_{H3} \\ z_{H3} \end{bmatrix}
$$

$$
\sigma \begin{bmatrix} x_N \\ y_N \\ z_N \\ x_{H1} \\ y_{H1} \\ z_{H1} \\ x_{H2} \\ y_{H2} \\ z_{H2} \\ x_{H3} \\ y_{H3} \\ z_{H3} \end{bmatrix} = \begin{bmatrix} 1 & 0 & 0 & 0 & 0 & 0 & 0 & 0 & 0 & 0 & 0 & 0 \\ 0 & 1 & 0 & 0 & 0 & 0 & 0 & 0 & 0 & 0 & 0 & 0 \\ 0 & 0 & 1 & 0 & 0 & 0 & 0 & 0 & 0 & 0 & 0 & 0 \\ 0 & 0 & 0 & 1 & 0 & 0 & 0 & 0 & 0 & 0 & 0 & 0 \\ 0 & 0 & 0 & 0 & 1 & 0 & 0 & 0 & 0 & 0 & 0 & 0 \\ 0 & 0 & 0 & 0 & 0 & 1 & 0 & 0 & 0 & 0 & 0 & 0 \\ 0 & 0 & 0 & 0 & 0 & 0 & 0 & 0 & 0 & -1 & 0 & 0 \\ 0 & 0 & 0 & 0 & 0 & 0 & 0 & 0 & 0 & 0 & -1 & 0 \\ 0 & 0 & 0 & 0 & 0 & 0 & 0 & 0 & 0 & 0 & 0 & 1 \\ 0 & 0 & 0 & 0 & 0 & 0 & -1 & 0 & 0 & 0 & 0 & 0 \\ 0 & 0 & 0 & 0 & 0 & 0 & 0 & -1 & 0 & 0 & 0 & 0 \\ 0 & 0 & 0 & 0 & 0 & 0 & 0 & 0 & 1 & 0 & 0 & 0 \end{bmatrix} \begin{bmatrix} x_N \\ y_N \\ z_N \\ x_{H1} \\ y_{H1} \\ z_{H1} \\ x_{H2} \\ y_{H2} \\ z_{H2} \\ x_{H3} \\ y_{H3} \\ z_{H3} \end{bmatrix}
$$

13.27. By multiplying the corresponding characters together, one can show that the following relationships exist for the D_2 point group:

$A_1 \times A_1 = A_1$ $\qquad A_1 \times B_1 = B_1$ $\qquad A_1 \times B_2 = B_2$ $\qquad A_1 \times B_3 = B_3$

$B_1 \times B_1 = A_1$ $\qquad B_1 \times B_2 = B_3$ $\qquad B_1 \times B_3 = B_2$ $\qquad B_2 \times B_2 = A_1$

$B_2 \times B_3 = B_1$ $\qquad B_3 \times B_3 = A_1$

Therefore, the closure requirement is satisfied.

13.29. The specific solution depends on which pair of irreducible representations you select. The following illustrates one pair for each point group.

(a) For C_2: $A \times B = (1 \cdot 1 \cdot 1 + 1 \cdot 1 \cdot -1) = 0$

(b) For C_{2v}: $A_1 \times B_1 = (1 \cdot 1 \cdot 1 + 1 \cdot 1 \cdot -1 + 1 \cdot 1 \cdot 1 + 1 \cdot 1 \cdot -1) = 0$

(c) For D_{2h}: $A_g \times A_u = (1 \cdot 1 \cdot 1 + 1 \cdot 1 \cdot 1 + 1 \cdot 1 \cdot 1 + 1 \cdot 1 \cdot 1 + 1 \cdot 1 \cdot -1 + 1 \cdot 1 \cdot -1 + 1 \cdot 1 \cdot -1 + 1 \cdot 1 \cdot -1) = 0$

(d) For O_h: $T_{1u} \times T_{2u} = (1 \cdot 3 \cdot 3 + 8 \cdot 0 \cdot 0 + 3 \cdot -1 \cdot -1 + 6 \cdot 1 \cdot -1 + 6 \cdot -1 \cdot 1 + 1 \cdot -3 \cdot -3 + 8 \cdot 0 \cdot 0$

$$+ 3 \cdot 1 \cdot 1 + 6 \cdot -1 \cdot 1 + 6 \cdot 1 \cdot -1) = (9 + 0 + 3 - 6 - 6 + 9 + 0 + 3 - 6 - 6) = 0$$

(e) For T_d: $E \times T_1 = (1 \cdot 2 \cdot 3 + 8 \cdot -1 \cdot 0 + 3 \cdot 2 \cdot -1 + 6 \cdot 0 \cdot 1 + 6 \cdot 0 \cdot -1) = 6 + 0 - 6 + 0 + 0 = 0$

Since all of the direct products equal zero, the irreducible representations are orthogonal. You should convince yourself that any two irreducible representations are orthogonal, not just these pairs.

13.31. Irreducible representations from different point groups will have different symmetry classes and orders. Therefore, the GOT can't be applied properly and the concept of orthogonality or normality is irrelevant.

13.33. In the exercise, the character for identity is given as 7. For the additional characters, we can follow Example 13.10 and use the formulas in the $R_h(3)$ character table, using the fact that j must equal 3. We note that f orbitals have different phases when we reflect through the origin, so we should use the "u" set of formulas (like the p orbitals in Example 13.10). We have:

$$\chi(E) = 2j + 1 = 7$$

$$\chi(C_3) = 1 + 2\cos(120°) + 2\cos(240°) + 2\cos(360°) = 1$$

$$\chi(C_2) = 1 + 2\cos(180°) + 2\cos(360°) + 2\cos(540°) = -1$$

$$\chi(C_4) = 1 + 2\cos(90°) + 2\cos(180°) + 2\cos(270°) = -1$$

$$\chi(C_2) = 1 + 2\cos(180°) + 2\cos(360°) + 2\cos(540°) = -1$$

$$\chi(i) = -(2j + 1) = -7$$

$$\chi(S_6) = -1 + 2\cos(60°) - 2\cos(120°) + 2\cos(180°) = -1$$

$$\chi(\sigma_h) = -(-1)^3 = 1$$

$$\chi(S_4) = -1 + 2\cos(90°) - 2\cos(180°) + 2\cos(270°) = 1$$

$$\chi(\sigma_d) = -(-1)^3 = 1$$

Therefore, we have the following character set:

	E	$8\,C_3$	$3\,C_2$	$6\,C_4$	$6\,C_2'$	i	$8\,S_6$	$3\,\sigma_h$	$6\,S_4$	$6\,\sigma_d$
Γ	7	1	-1	-1	-1	-7	-1	1	1	1

We can reduce this into its irreducible representations by using the GOT:

$$a(A_{1g}) = \frac{1}{48}\big(7 \cdot 1 \cdot 1 + 1 \cdot 1 \cdot 8 + (-1) \cdot 1 \cdot 3 + (-1) \cdot 1 \cdot 6 + (-1) \cdot 1 \cdot 6 + (-7) \cdot 1 \cdot 1 + (-1) \cdot 1 \cdot 8 + 1 \cdot 1 \cdot 3$$

$$+ 1 \cdot 1 \cdot 6 + 1 \cdot 1 \cdot 6\big) = 0 \; A_{1g}$$

$$a(A_{2g}) = \frac{1}{48}\big(7\cdot 1\cdot 1 + 1\cdot 1\cdot 8 + (-1)\cdot 1\cdot 3 + (-1)\cdot(-1)\cdot 6 + (-1)\cdot(-1)\cdot 6 + (-7)\cdot 1\cdot 1 + (-1)\cdot 1\cdot 8 + 1\cdot 1\cdot 3$$
$$+ 1\cdot(-1)\cdot 6 + 1\cdot(-1)\cdot 6\big) = 0 \; A_{2g}$$

$$a(E_{g}) = \frac{1}{48}\big(7\cdot 2\cdot 1 + 1\cdot(-1)\cdot 8 + (-1)\cdot 2\cdot 3 + (-1)\cdot 0\cdot 6 + (-1)\cdot 0\cdot 6 + (-7)\cdot 2\cdot 1 + (-1)\cdot(-1)\cdot 8 + 1\cdot 2\cdot 3$$
$$+ 1\cdot 0\cdot 6 + 1\cdot 0\cdot 6\big) = 0 \; E_{g}$$

$$a(T_{1g}) = \frac{1}{48}\big(7\cdot 3\cdot 1 + 1\cdot 0\cdot 8 + (-1)\cdot(-1)\cdot 3 + (-1)\cdot 1\cdot 6 + (-1)\cdot(-1)\cdot 6 + (-7)\cdot 3\cdot 1 + (-1)\cdot 0\cdot 8 + 1\cdot(-1)\cdot 3$$
$$+ 1\cdot 1\cdot 6 + 1\cdot(-1)\cdot 6\big) = 0 \; T_{1g}$$

$$a(T_{2g}) = \frac{1}{48}\big(7\cdot 3\cdot 1 + 1\cdot 0\cdot 8 + (-1)\cdot(-1)\cdot 3 + (-1)\cdot(-1)\cdot 6 + (-1)\cdot 1\cdot 6 + (-7)\cdot 3\cdot 1 + (-1)\cdot 0\cdot 8 + 1\cdot(-1)\cdot 3$$
$$+ 1\cdot(-1)\cdot 6 + 1\cdot 1\cdot 6\big) = 0 \; T_{2g}$$

$$a(A_{1u}) = \frac{1}{48}\big(7\cdot 1\cdot 1 + 1\cdot 1\cdot 8 + (-1)\cdot 1\cdot 3 + (-1)\cdot 1\cdot 6 + (-1)\cdot 1\cdot 6 + (-7)\cdot(-1)\cdot 1 + (-1)\cdot(-1)\cdot 8 + 1\cdot(-1)\cdot 3$$
$$+ 1\cdot(-1)\cdot 6 + 1\cdot(-1)\cdot 6\big) = 0 \; A_{1u}$$

$$a(A_{2u}) = \frac{1}{48}\big(7\cdot 1\cdot 1 + 1\cdot 1\cdot 8 + (-1)\cdot 1\cdot 3 + (-1)\cdot(-1)\cdot 6 + (-1)\cdot(-1)\cdot 6 + (-7)\cdot(-1)\cdot 1 + (-1)\cdot(-1)\cdot 8$$
$$+ 1\cdot(-1)\cdot 3 + 1\cdot 1\cdot 6 + 1\cdot 1\cdot 6\big) = 1 \; A_{2u}$$

$$a(E_{u}) = \frac{1}{48}\big(7\cdot 2\cdot 1 + 1\cdot(-1)\cdot 8 + (-1)\cdot 2\cdot 3 + (-1)\cdot 0\cdot 6 + (-1)\cdot 0\cdot 6 + (-7)\cdot(-2)\cdot 1 + (-1)\cdot 1\cdot 8 + 1\cdot(-2)\cdot 3$$
$$+ 1\cdot 0\cdot 6 + 1\cdot 0\cdot 6\big) = 0 \; E_{u}$$

$$a(T_{1u}) = \frac{1}{48}\big(7\cdot 3\cdot 1 + 1\cdot 0\cdot 8 + (-1)\cdot(-1)\cdot 3 + (-1)\cdot 1\cdot 6 + (-1)\cdot(-1)\cdot 6 + (-7)\cdot(-3)\cdot 1 + (-1)\cdot 0\cdot 8 + 1\cdot 1\cdot 3$$
$$+ 1\cdot 1\cdot 6 + 1\cdot(-1)\cdot 6\big) = 1 \; T_{1u}$$

$$a(T_{2u}) = \frac{1}{48}\big(7\cdot 3\cdot 1 + 1\cdot 0\cdot 8 + (-1)\cdot(-1)\cdot 3 + (-1)\cdot(-1)\cdot 6 + (-1)\cdot 1\cdot 6 + (-7)\cdot(-3)\cdot 1 + (-1)\cdot 0\cdot 8 + 1\cdot 1\cdot 3$$
$$+ 1\cdot 1\cdot 6 + 1\cdot(-1)\cdot 6\big) = 1 \; T_{2u}$$

Thus, the irreducible representation that describes the seven f orbitals in octahedral symmetry is $A_{2u} + T_{1u} + T_{2u}$.

13.35. In the case where the position of symmetry is the point $x = \pi/2$, the symmetry elements are E, C_2 (the $x = \pi/2$ axis), and 2 σ's (the xy and the yz planes).

13.37. According to the scheme in Figure 13.14, ethylene (C_2H_4) has D_{2h} point group symmetry. With the center of inversion being in the middle of the molecule, we can see that the inversion operation changes the sign on the wavefunction; therefore, the character of i should be –1. According to how the axes are defined, the xz plane is the plane of the page, and since the orbitals are being reflected onto themselves, the character of σ(xz) should be +1. Finally, the yz plane cuts through the plane of the page between the two orbitals. Reflecting the orbitals through that plane reflects them onto an orbital of the same phase, so the character of σ(yz) should also be +1. This is enough information to indicate that these orbitals have the B_{3u} irreducible representation in the D_{2h} point group.

13.39. (a) $a(A) = \dfrac{1}{2}(5 \cdot 1 \cdot 1 + 1 \cdot 1 \cdot 1) = 3A$

$a(B) = \dfrac{1}{2}(5 \cdot 1 \cdot 1 + 1 \cdot 1 \cdot (-1)) = 2B$

Therefore, in this case, $\Gamma = 3A + 2B$.

(b) $a(A_1) = \dfrac{1}{6}(6 \cdot 1 \cdot 1 + 0 \cdot 2 \cdot 1 + 0 \cdot 3 \cdot 1) = 1A_1$

$a(A_2) = \dfrac{1}{6}(6 \cdot 1 \cdot 1 + 0 \cdot 2 \cdot 1 + 0 \cdot 3 \cdot (-1)) = 1A_2$

$a(E) = \dfrac{1}{6}(6 \cdot 1 \cdot 2 + 0 \cdot 2 \cdot (-1) + 0 \cdot 3 \cdot 0) = 2E$

Therefore, in this case, $\Gamma = A_1 + A_2 + 2E$.

(c) $a(A_1) = \dfrac{1}{8}(6 \cdot 1 \cdot 1 + (-2) \cdot 2 \cdot 1 + 2 \cdot 1 \cdot 1 + 2 \cdot 2 \cdot 1 + -4 \cdot 2 \cdot 1) = 0A_1$

$a(A_2) = \dfrac{1}{8}(6 \cdot 1 \cdot 1 + (-2) \cdot 2 \cdot 1 + 2 \cdot 1 \cdot 1 + 2 \cdot 2 \cdot (-1) + -4 \cdot 2 \cdot (-1)) = 1A_2$

$a(B_1) = \dfrac{1}{8}(6 \cdot 1 \cdot 1 + (-2) \cdot 2 \cdot (-1) + 2 \cdot 1 \cdot 1 + 2 \cdot 2 \cdot 1 + -4 \cdot 2 \cdot (-1)) = 3B_1$

$a(B_2) = \dfrac{1}{8}(6 \cdot 1 \cdot 1 + (-2) \cdot 2 \cdot (-1) + 2 \cdot 1 \cdot 1 + 2 \cdot 2 \cdot (-1) + -4 \cdot 2 \cdot 1) = 0B_2$

$a(E) = \dfrac{1}{8}(6 \cdot 1 \cdot 2 + (-2) \cdot 2 \cdot 0 + 2 \cdot 1 \cdot (-2) + 2 \cdot 2 \cdot 0 + -4 \cdot 2 \cdot 0) = 1E$

Therefore, in this case, $\Gamma = A_2 + 3B_1 + E$.

(d) $a(A_1) = \dfrac{1}{24}(7 \cdot 1 \cdot 1 + (-2) \cdot 8 \cdot 1 + 3 \cdot 3 \cdot 1 + 1 \cdot 6 \cdot 1 + (-1) \cdot 6 \cdot 1) = 0A_1$

$a(A_2) = \dfrac{1}{24}(7 \cdot 1 \cdot 1 + (-2) \cdot 8 \cdot 1 + 3 \cdot 3 \cdot 1 + 1 \cdot 6 \cdot (-1) + (-1) \cdot 6 \cdot (-1)) = 0A_2$

$a(E) = \dfrac{1}{24}(7 \cdot 1 \cdot 2 + (-2) \cdot 8 \cdot (-1) + 3 \cdot 3 \cdot 2 + 1 \cdot 6 \cdot 0 + (-1) \cdot 6 \cdot 0) = 2E$

$a(T_1) = \dfrac{1}{24}(7 \cdot 1 \cdot 3 + (-2) \cdot 8 \cdot 0 + 3 \cdot 3 \cdot (-1) + 1 \cdot 6 \cdot 1 + (-1) \cdot 6 \cdot (-1)) = 1T_1$

$a(T_2) = \dfrac{1}{24}(7 \cdot 1 \cdot 3 + (-2) \cdot 8 \cdot 0 + 3 \cdot 3 \cdot (-1) + 1 \cdot 6 \cdot (-1) + (-1) \cdot 6 \cdot 1) = 0T_2$

Therefore, in this case, $\Gamma = 2E + T_1$.

13.41. In order for a product of functions to be non-zero, the product of the irreducible representations must contain the all-symmetric irreducible representation, usually labeled A_1 (or A or A' or A_1' or A_{1g}). If the product of irreducible representations does not contain A_1, then the integral must be exactly zero.

(a) In C_{3v}, the product $A_1 \times A_2$ yields the character set (1, 1, -1), which is the A_2 irreducible representation. Thus, $A_1 \times A_2 = A_2$. Thus, an integral having these irreducible representations must equal zero.

(b) In C_{6v}, the product $E_1 \times E_2$ yields the character set (4, -1, 1, -4, 0, 0). Using the GOT to reduce this representation, we find that it equals $B_1 + B_2 + E_1$. Thus, an integral having these irreducible representations must equal zero.

(c) In D_{3h}, the product $A_2' \times A_1'' \times E''$ yields the character set (2, -1, 0, 2, -1, 0), which is the E' irreducible representation. Thus, an integral having these irreducible representations must equal zero.

(d) In D_{6h}, the product $B_{2g} \times B_{2u}$ yields the character set (1, 1, 1, 1, 1, 1, -1, -1, -1, -1, -1, -1), which is the A_{1u} irreducible representation. Thus, an integral having these irreducible representations must equal zero.

(e) In D_{6h}, the product $B_{1g} \times B_{1g}$ yields the character set (1, 1, 1, 1, 1, 1, 1, 1, 1, 1, 1, 1), which is the A_{1g} irreducible representation. Thus, an integral having these irreducible representations may be non-zero.

(f) In T_d, the product $E \times T_1$ yields the character set (6, 0, -2, 0, 0). Using the GOT to reduce this representation, we find that it equals $T_1 + T_2$. Thus, an integral having these irreducible representations must equal zero.

(g) In T_d, the product $T_2 \times T_2$ yields the character set (9, 0, 1, 1, 1). Using the GOT to reduce this representation, we find that it equals $A_1 + E + T_1 + T_2$. Thus, an integral having these irreducible representations may be non-zero.

13.43. Because s orbitals are spherically symmetric, they have the $D_g^{(0)}$ irreducible representation in the $R_h(3)$ point group. If the operator has the irreducible representation $D_u^{(1)}$, the integral of interest has an overall symmetry given by the product $D_g^{(0)} \times D_u^{(1)} \times D_g^{(0)}$. The product of these three irreducible representations is equal to $D_u^{(1)}$. Since this is not the all-symmetric irreducible representation (represented in this point group by $D_g^{(0)}$), the integral is exactly zero.

13.45. H_2S has C_{2v} symmetry. To determine the symmetry-adapted linear combination for the molecular orbitals of H_2S, we follow the scheme from Section 13.9 but using only the valence orbitals for H and S:

	$1s_{H1}$	$1s_{H2}$	$3s_S$	$3p_{x,S}$	$3p_{y,S}$	$3p_{z,S}$
E	$1s_{H1}$	$1s_{H2}$	$3s_S$	$3p_{x,S}$	$3p_{y,S}$	$3p_{z,S}$
C_2	$1s_{H2}$	$1s_{H1}$	$3s_S$	$-3p_{x,S}$	$-3p_{y,S}$	$3p_{z,S}$
σ	$1s_{H1}$	$1s_{H2}$	$3s_S$	$3p_{x,S}$	$-3p_{y,S}$	$3p_{z,S}$
σ'	$1s_{H2}$	$1s_{H1}$	$3s_S$	$-3p_{x,S}$	$3p_{y,S}$	$3p_{z,S}$

The combinations for the A_1 wavefunctions are determined by determined by multiplying each row of functions by the character of the row's symmetry element, adding each column, and dividing by the order of the group (4, in this case). We get:

$$\Psi = \frac{1}{4}\left(1s_{H1} + 1s_{H2} + 1s_{H1} + 1s_{H2}\right) = \frac{1}{2}\left(1s_{H1} + 1s_{H2}\right)$$

$$\Psi = \frac{1}{4}\left(1s_{H2} + 1s_{H1} + 1s_{H2} + 1s_{H1}\right) = \frac{1}{2}\left(1s_{H1} + 1s_{H2}\right)$$

$$\Psi = \frac{1}{4}\left(3s_S + 3s_S + 3s_S + 3s_S\right) = 3s_S$$

$$\Psi = \frac{1}{4}\left(3p_{x,S} - 3p_{x,S} + 3p_{x,S} - 3p_{x,S}\right) = 0$$

$$\Psi = \frac{1}{4}\left(3p_{y,S} - 3p_{y,S} - 3p_{y,S} + 3p_{y,S}\right) = 0$$

$$\Psi = \frac{1}{4}\left(3p_{z,S} + 3p_{z,S} + 3p_{z,S} + 3p_{z,S}\right) = 3p_{z,S}$$

The only unique wavefunctions we get are $\Psi = \frac{1}{2}(1s_{H1} + 1s_{H2})$, $\Psi = 3s_S$, and $\Psi = 3p_{z,S}$. For the A_2 set of wavefunctions:

$$\Psi = \frac{1}{4}\left(1s_{H1} + 1s_{H2} - 1s_{H1} - 1s_{H2}\right) = 0$$

$$\Psi = \frac{1}{4}\left(1s_{H2} + 1s_{H1} - 1s_{H2} - 1s_{H1}\right) = 0$$

$$\Psi = \frac{1}{4}\left(3s_S + 3s_S - 3s_S - 3s_S\right) = 0$$

$$\Psi = \frac{1}{4}\left(3p_{x,S} - 3p_{x,S} - 3p_{x,S} + 3p_{x,S}\right) = 0$$

$$\Psi = \frac{1}{4}\left(3p_{y,S} - 3p_{y,S} + 3p_{y,S} - 3p_{y,S}\right) = 0$$

$$\Psi = \frac{1}{4}\left(3p_{z,S} + 3p_{z,S} - 3p_{z,S} - 3p_{z,S}\right) = 0$$

Thus, there are no SALC wavefunctions that belong to the A_2 irreducible representation. For the B_1 set of wavefunctions:

$$\Psi = \frac{1}{4}\left(1s_{H1} - 1s_{H2} + 1s_{H1} - 1s_{H2}\right) = 0$$

$$\Psi = \frac{1}{4}\left(1s_{H2} - 1s_{H1} + 1s_{H2} - 1s_{H1}\right) = \frac{1}{2}\left(1s_{H2} - 1s_{H1}\right)$$

$$\Psi = \frac{1}{4}\left(3s_S - 3s_S + 3s_S - 3s_S\right) = 0$$

$$\Psi = \frac{1}{4}\left(3p_{x,S} + 3p_{x,S} + 3p_{x,S} + 3p_{x,S}\right) = 3p_{x,S}$$

$$\Psi = \frac{1}{4}\left(3p_{y,S} + 3p_{y,S} - 3p_{y,S} - 3p_{y,S}\right) = 0$$

$$\Psi = \frac{1}{4}\left(3p_{z,S} - 3p_{z,S} + 3p_{z,S} - 3p_{z,S}\right) = 0$$

Thus, there are two SALC wavefunctions that have B_1 symmetry: $\Psi = \frac{1}{2}(1s_{H2} - 1s_{H1})$ and $\Psi = 3p_{x,S}$. For the B_2 set of wavefunctions:

$$\Psi = \frac{1}{4}\left(1s_{H1} - 1s_{H2} - 1s_{H1} + 1s_{H2}\right) = 0$$

$$\Psi = \frac{1}{4}\left(1s_{H1} - 1s_{H2} - 1s_{H1} + 1s_{H2}\right) = 0$$

$$\Psi = \frac{1}{4}\left(3s_S - 3s_S - 3s_S + 3s_S\right) = 0$$

$$\Psi = \frac{1}{4}\left(3p_{x,S} + 3p_{x,S} - 3p_{x,S} - 3p_{x,S}\right) = 0$$

$$\Psi = \frac{1}{4}\left(3p_{y,S} + 3p_{y,S} + 3p_{y,S} + 3p_{y,S}\right) = 3p_{y,S}$$

$$\Psi = \frac{1}{4}\left(3p_{z,S} - 3p_{z,S} - 3p_{z,S} + 3p_{z,S}\right) = 0$$

Thus, there is one B_2 wavefunction: $\Psi = 3p_{y,S}$. This gives us a full set of six molecular orbitals having the proper irreducible representations for this molecule.

13.47. For many molecules, a reasonably good approximation for a molecular orbital may in fact be an atomic orbital, especially if that particular atomic orbital is not involved in bonding. Many core orbitals – orbitals not part of the valence shell – do not participate in bonding, so by themselves they may be good representatives of molecular orbitals.

13.49. We will be using four $1s$ orbitals from the hydrogen atoms in CH_4, along with the $1s$, $2s$, $2p_x$, $2p_y$, and $2p_z$ orbitals of the carbon atom. This gives us a total of 9 atomic orbitals to construct the SALCs, and ultimately we should get 9 independent combinations for the molecular orbitals.

13.51. As indicated in Equation 13.12, the "first excited state" of H_2 is actually a set of three wavefunctions that have the same spatial part but different spin parts. Although we assume that the spin part of a wavefunction will not affect its energy, in reality the different spin functions will have a tiny effect on the energy of the overall wavefunction. A detailed enough spectrum will show that the "first excited state" will have three separate lines, not one.

13.53. $\int \left(\frac{1}{\sqrt{2}}(s+p_z)\right)^* \frac{1}{\sqrt{2}}(s-p_z)d\tau = \frac{1}{2}\left[\int s^* s\, d\tau - \int s^* p_z d\tau + \int p_z^* s\, d\tau - \int p_z^* p_z d\tau\right]$

The individual atomic orbitals are themselves normalized and orthogonal to each other. Therefore, the first and last integrals are 1 and the second and third integrals are 0. Subsituting:

$= \frac{1}{2}(1 - 0 + 0 - 1) = 0$. Thus, the two hybrid orbitals are orthogonal.

13.55. If the CH_3^+ ion is roughly trigonal planar, then the carbon atom should have sp^2 hybrid orbitals making bonds to the hydrogens, as this is the only hybridization scheme that yields the appropriate directions of bonds.

13.57. The character for E is 5, as all five orbitals operate onto themselves. The character for C_3 would be 2. The character for C_2 would be 1, for σ_h would be 3, for S_3 would be 0, and for σ_v would be 3. Therefore, the set of characters for the five sp^2d hybrid orbitals would be (5, 2, 1, 3, 0, 3). Using the GOT, this representation reduces to $2A_1' + A_2'' + E'$.

13.59. The two orbitals are not orthogonal because they are on different atoms. Hence, there is no requirement that they be orthogonal. (In fact, if they were, no bond would be formed!)

13.61. The nitrogen atom not only must accommodate three bonds to hydrogen atoms, but also a lone electron pair. Thus, the nitrogen atom needs four hybrid orbitals: sp^3 orbitals.

CHAPTER 14. ROTATIONAL AND VIBRATIONAL SPECTROSCOPY

14.1. For a linear molecule, rotation about the molecular axis isn't recognized as a true rotation. Even if it were, the moment of inertia about that axis is almost identically zero, suggesting that only a negligible amount of energy is needed to promote rotation about that axis. A useful spectrum couldn't ever be measured.

14.3. Using the equation $c = \lambda \nu$:
(a) 2.9979×10^8 m/s $= (1.00$ m$)\nu$ $\nu = 3.00 \times 10^8$ s^{-1}
(b) 2.9979×10^8 m/s $= (4.77 \times 10^{-5}$ m$)\nu$ $\nu = 6.28 \times 10^{12}$ s^{-1}
(c) 7894 A $= 7.894 \times 10^{-7}$ m 2.9979×10^8 m/s $= (7.894 \times 10^{-7}$ m$)\nu$
$\nu = 3.798 \times 10^{14}$ s^{-1}
(d) 2.9979×10^8 m/s $= (1.903 \times 10^3$ m$)\nu$ $\nu = 1.575 \times 10^5$ s^{-1}

14.5. 2.9979×10^8 m/s $= \lambda(8.041 \times 10^{12}$ s$^{-1})$ $\lambda = 3.728 \times 10^{-5}$ m
$E = h\nu$ $E = (6.626 \times 10^{-34}$ J· s$)(8.041 \times 10^{12}$ s$^{-1})$ $E = 5.328 \times 10^{-21}$ J
The speed of the photon is the same as any other photon's speed (in vacuum): 2.9979×10^8 m/s.

14.7. Since there are 1,000,000 micrometers in 1 meter and 100 m^{-1} in 1 cm^{-1}, you can show that for a photon of light having x cm^{-1} and y μm, $x \times y = 10{,}000$.

14.9. (a) prolate symmetric top (b) spherical top (c) spherical top (d) asymmetric top (e) asymmetric top (f) asymmetric top (g) oblate symmetric top (h) linear (i) linear.

14.11. Both SF_6 and UF_6 are octahedral, spherical-top molecules. Therefore, their moments of inertia are independent of the axis of rotation. Therefore, we can choose a simple-to-envision rotation and use that to calculate the moments of inertia, and from that, the value of B. Let us select rotation of the molecule along one of the F-S-F or F-U-F axes. That way, the three atoms along the axis do not contribute to the moment of inertia because they are not rotating. What is rotating are four F atoms a particular distance away from the rotational axis. The moments of inertia can be calculated simply, by considering it as four atoms rotating about an axis. For SF_6:
$I = 4 \times (0.0190$ kg$/6.022 \times 10^{23})(1.564 \times 10^{-10}$ m$)^2 = 3.087 \times 10^{-45}$ kg· m^2
For UF_6: $I = 4 \times (0.0190$ kg$/6.022 \times 10^{23})(1.996 \times 10^{-10}$ m$)^2 = 5.028 \times 10^{-45}$ kg· m^2
Now to calculate the values of B:
$$B(SF_6) = \frac{(6.626 \times 10^{-34} \text{ J} \cdot \text{s})^2}{(2\pi)^2 2(3.087 \times 10^{-45} \text{ kg} \cdot \text{m}^2)} = 1.80 \times 10^{-24} \text{ J}$$
$$B(UF_6) = \frac{(6.626 \times 10^{-34} \text{ J} \cdot \text{s})^2}{(2\pi)^2 2(5.028 \times 10^{-45} \text{ kg} \cdot \text{m}^2)} = 1.10 \times 10^{-24} \text{ J}$$
When you consider that the central atoms are so different in mass, these values of B are not that much different from each other!

14.13. There are $2J + 1$ M_J levels that are degenerate. Also, for each of these levels, the positive and negative values of K that have the same magnitude also yield the same energy, because energy is dependent on K^2. Therefore, for most energy levels, the degeneracy is $2 \times (2J + 1)$. When $K = 0$, there's not another value of K that yields the same energy, since $-0 = 0$.

Therefore, the degeneracy of those energy levels is dictated solely by the various M_J values – $2J$ + 1 of them.

14.15. Ethane is a prolate symmetric top with the two lower rotational constants being the same:

$$A = \frac{(6.626 \times 10^{-34} \text{ J} \cdot \text{s})^2}{(2\pi)^2 2(1.075 \times 10^{-46} \text{ kg} \cdot \text{m}^2)} = 5.173 \times 10^{-23} \text{ J}$$

$$B = C = \frac{(6.626 \times 10^{-34} \text{ J} \cdot \text{s})^2}{(2\pi)^2 2(4.200 \times 10^{-46} \text{ kg} \cdot \text{m}^2)} = 1.324 \times 10^{-23} \text{ J}$$

The lowest rotational energy level has $E = 0$. The second energy level comes from the $J = 1$ value of the lower of the two rotational constants:

$$E = BJ(J+1) = 1(1+1) \cdot 1.324 \times 10^{-23} \text{ J} = 2.648 \times 10^{-23} \text{ J}$$

The third energy level comes from the $J = 2$ value of the lower of the two rotational constants:

$$E = BJ(J+1) = 2(2+1) \cdot 1.324 \times 10^{-23} \text{ J} = 7.944 \times 10^{-23} \text{ J}$$

The fourth energy level comes from the $J = 1$ value of the higher of the two constants:

$$E = AJ(J+1) = 1(1+1) \cdot 5.173 \times 10^{-23} \text{ J} = 1.035 \times 10^{-22} \text{ J}$$

Finally, the fifth energy level comes from the $J = 3$ value of the lower of the two constants:

$$E = BJ(J+1) = 3(3+1) \cdot 1.324 \times 10^{-23} \text{ J} = 1.589 \times 10^{-22} \text{ J}$$

14.17. (a) allowed (b) not allowed (c) not allowed (d) not allowed (e) not allowed (f) allowed (g) not allowed (h) allowed

14.19. If the rotational spectrum consists of lines spaced by 0.114 cm^{-1}, this value represents $2B$, so $B = 0.057$ cm^{-1}. Calculating the reduced mass of ICl:

$$\mu = \frac{(127)(35)}{(127+35)} \text{ g} \times \frac{1 \text{ kg}}{1000 \text{ g}} \div 6.022 \times 10^{23} = 4.556 \times 10^{-26} \text{ kg}$$

Using the wavenumber forms of the equation for the rotational constant:

$$B = \frac{h}{8\pi^2(\mu r^2)c} = \frac{6.626 \times 10^{-34} \text{ J} \cdot \text{s}}{(8\pi^2)(4.556 \times 10^{-26} \text{ kg})r^2(2.9979 \times 10^{10} \text{ cm/s})} = 0.057 \text{ cm}^{-1}$$

Solving for r^2, then for r:

$$r^2 = \frac{6.626 \times 10^{-34} \text{ J} \cdot \text{s}}{(8\pi^2)(4.556 \times 10^{-26} \text{ kg})(0.057 \text{ cm}^{-1})(2.9979 \times 10^{10} \text{ cm/s})} = 1.078 \times 10^{-19} \text{ m}^2$$

$r = 3.28 \times 10^{-10}$ m = 3.28 A.

14.21. The easiest way to solve this question is to convert the B value into GHz units, then use the fact that the first four absorptions will appear at $2B$, $4B$, $6B$, and $8B$. If the rotational spectrum consists of lines spaced by 15.026 cm^{-1}, this value represents $2B$, so $B = 7.513$ cm^{-1}. Converting this into gigahertz units:

$$\lambda = \frac{1}{\tilde{\nu}} = \frac{1}{7.513 \text{ cm}^{-1}} = 0.1331 \text{ cm} = 0.001331 \text{ m} \qquad \text{Calculating the frequency:}$$

$$c = \lambda \nu \qquad 2.9979 \times 10^8 \text{ m/s} = (0.001331 \text{ m})\nu \qquad \nu = 2.252 \times 10^{11} \text{ s}^{-1} = 225.2 \text{ GHz}$$

Therefore, in GHz units, $B = 225.2$ GHz. Therefore, the first four lines of the rotational spectrum should appear at 450.4 GHz, 900.8 GHz, 1351.2 GHz, and 1801.6 GHz.

14.23. Calculating the reduced mass of HS:

$$\mu = \frac{(32)(1)}{(32+1)} g \times \frac{1 \, kg}{1000 \, g} \div 6.022 \times 10^{23} = 1.610 \times 10^{-27} \, kg$$

This lets us calculate B for the diatomic molecule:

$$B = \frac{(6.626 \times 10^{-34} \, J \cdot s)^2}{(2\pi)^2 2(1.610 \times 10^{-27} \, kg)(1.40 \times 10^{-10} \, m)^2} = 1.762 \times 10^{-22} \, J$$

Knowing that J_{max} is 8, we can estimate the sample temperature:

$$8 \approx \left(\frac{(1.381 \times 10^{-23} \, J/K)T}{2(1.762 \times 10^{-22} \, J)} \right)^{1/2} \qquad \text{Solve for } T: \ T \approx 1600 \, K$$

14.25. From Table 14.2, we find that B (HCl) $= 10.59$ cm^{-1}. Since $B = \dfrac{\hbar^2}{2I} = \dfrac{\hbar^2}{2\mu r^2}$, we can

determine B (DCl) by determining the change in the reduced masses (everything else remains the same):

$$\mu \, (HCl) = \frac{(1)(35.5)}{(1+35.5)} g = 0.9726 \, g \qquad \qquad \mu \, (DCl) = \frac{(2)(35.5)}{(2+35.5)} g = 1.8933 \, g$$

(Here we are simply using the atomic weight of Cl, rather than a particular isotopic weight.) Since the reduced mass is in the denominator of the expression for B, we can set up the following ratio:

$$\frac{B \, (DCl)}{B \, (HCl)} = \frac{1/\mu \, (DCl)}{1/\mu \, (HCl)} = \frac{\mu \, (HCl)}{\mu \, (DCl)} \qquad \text{Substituting and solving for } B \text{ (DCl):}$$

$$\frac{B \, (DCl)}{10.59 \, cm^{-1}} = \frac{0.9726 \, g}{1.8933 \, g} \qquad \qquad B \, (DCl) = 5.44 \, cm^{-1}.$$

14.27. First, at such high values of rotational quantum number, we expect that centrifugal distortions will make the energy levels (and hence the energy difference) deviate quite a bit from those of a rigid rotor. Second, as a nonpolar molecule, diatomic iodine wouldn't show a pure rotational spectrum.

14.29. Using equation 14.27:

$$D_J \approx \frac{4B^3}{\tilde{\nu}^2} = \frac{4(60.80 \, cm^{-1})^3}{(4320 \, cm^{-1})^2} = \frac{8.990 \times 10^5 \, cm^{-3}}{1.8662 \times 10^7 \, cm^{-2}} = 0.04817 \, cm^{-1}$$

This compares to 4.64×10^{-2} cm^{-1}, which is not far off (about 0.2%).

14.31. The total number of normal modes equals the number of vibrational degrees of freedom. (We should differentiate the *total* number of normal modes from the total number of *distinct* vibrational frequencies, which may be different due to degeneracies.) Therefore, the total number of normal modes for the molecules in exercise 14.30 are (a) 1; (b) 3; (c) 174; (d) 63; (e) 48; (f) 7; and (g) 6.

14.33. Starting with the equation $\dfrac{1}{\mu} = \dfrac{1}{m_1} + \dfrac{1}{m_2}$, we multiply the first fraction on the right by m_2/m_2 and the second fraction on the right by m_1/m_1:

$\dfrac{1}{\mu} = \dfrac{m_2}{m_1 m_2} + \dfrac{m_1}{m_1 m_2} = \dfrac{m_1 + m_2}{m_1 m_2}$. Now we can take the reciprocal of both sides of the equation to

get an equivalent expression for the reduced mass: $\mu = \dfrac{m_1 m_2}{m_1 + m_2}$.

14.35. First, we need the reduced masses of the C=O and C=S bonds:

For C = O: $\mu = \dfrac{(12)(18)}{12 + 18}$ g = 7.2 g   For C = S: $\mu = \dfrac{(12)(32)}{12 + 32}$ g = 8.73 g

Now we use the square root of the ratio of the two reduced masses to predict the vibrational frequency of the C=S bond:

$\sqrt{\dfrac{7.2 \text{ g}}{8.73 \text{ g}}} = 0.9082 = \dfrac{\tilde{\nu}\,(C = S)}{1338 \text{ cm}^{-1}}$ $\tilde{\nu}\,(C = S) = 1215 \text{ cm}^{-1}$

Although we predict 1215 cm^{-1} for the frequency of the C=S vibration, it actually appears at 859 cm^{-1}. Thus, assuming that S is an "isotope" of O is not a good assumption.

14.37. Nitrogen is used because, as a nonpolar diatomic molecule, its single vibrational motion is IR-inactive. Oxygen gas and any noble gas (He, Ne, Ar, Kr) could also be used – but they are more expensive. Thus, using nitrogen gas as a purge gas means that the infrared spectrum is measuring the vibrations of the sample, not any IR-absorbing contaminant from the atmosphere.

14.39. Fundamental vibrations are the $v = 0 \rightarrow v = 1$ transitions. Overtone vibrations are the $v = 0 \rightarrow v = n$, where n is any number other than 1. Hot bands are the $v = n \rightarrow v = n + 1$, where n is any number other than 1 (that is, the vibrational transition starts in an excited vibrational state).

14.41. Starting with the expression $V = \dfrac{1}{2}kx^2$, we take the derivative of this expression with

respect to x twice: $\dfrac{\partial V}{\partial x} = kx$ $\dfrac{\partial^2 V}{\partial x^2} = k$, thus verifying the relationship. With regard to units, according to the original expression of Hooke's law:
(joules) = (N/m)(m)^2 J = N· m   J = J
Thus, the units are algebraically correct.

14.43. We need to go through a three-step process to determine a: first, use the data in Table 14.4 to calculate the "pure" anharmonicity constant x_e; second, use equation 14.40 to determine D_e; third, use equation 14.37 along with data given in the exercise to determine a.

For HF: $x_e = \dfrac{x_e \nu_e}{\nu_e} = \dfrac{90.07 \text{ cm}^{-1}}{4138.52 \text{ cm}^{-1}} = 0.02176$ Rearranging equation 14.40:

$$D_e = \frac{v_e}{4x_e} = \frac{4138.52\ \text{cm}^{-1}}{4(0.02176)} = 47{,}550\ \text{cm}^{-1}$$

And finally, converting the wavenumber value of D_e into joules:

$$a = \left(\frac{k}{2D_e}\right)^{1/2} = \left(\frac{965.1\ \text{N/m}}{2(47{,}550\ \text{cm}^{-1})} \cdot \frac{1}{2.9979\times10^{10}\ \text{cm/s}} \cdot \frac{1}{6.626\times10^{-34}\ \text{J}\cdot\text{s}}\right)^{1/2}$$

$a = 2.26\times10^{10}\ \text{m}^{-1} = 2.26\ \text{Å}^{-1}$.

For HBr: $\quad x_e = \dfrac{x_e v_e}{v_e} = \dfrac{45.21\ \text{cm}^{-1}}{2649.67\ \text{cm}^{-1}} = 0.01706 \quad$ Rearranging equation 14.40:

$$D_e = \frac{v_e}{4x_e} = \frac{2649.67\ \text{cm}^{-1}}{4(0.01706)} = 38{,}830\ \text{cm}^{-1}$$

And finally, converting the wavenumber value of D_e into joules:

$$a = \left(\frac{k}{2D_e}\right)^{1/2} = \left(\frac{411.5\ \text{N/m}}{2(38{,}830\ \text{cm}^{-1})} \cdot \frac{1}{2.9979\times10^{10}\ \text{cm/s}} \cdot \frac{1}{6.626\times10^{-34}\ \text{J}\cdot\text{s}}\right)^{1/2}$$

$a = 1.63\times10^{10}\ \text{m}^{-1} = 1.63\ \text{Å}^{-1}$.

Comparing these values with $1.87\ \text{Å}^{-1}$ for HCl, we find a trend that a decreases as the size of the halogen atom increases.

14.45. Using the fact that $D_e = D_0 + \frac{1}{2}hv$, we must first convert the given D_0 values into cm^{-1} units:

$$362\ \text{kJ/mol}\times\frac{1000\ \text{J}}{1\ \text{kJ}}\times\frac{1\ \text{mol}}{6.02\times10^{23}}\times\frac{1}{6.626\times10^{-34}\ \text{J}\cdot\text{s}}\times\frac{\text{s}}{2.9979\times10^{10}\ \text{cm}} = 30{,}270\ \text{cm}^{-1}$$

for HBr, and for CO:

$$1071\ \text{kJ/mol}\times\frac{1000\ \text{J}}{1\ \text{kJ}}\times\frac{1\ \text{mol}}{6.02\times10^{23}}\times\frac{1}{6.626\times10^{-34}\ \text{J}\cdot\text{s}}\times\frac{\text{s}}{2.9979\times10^{10}\ \text{cm}} = 89{,}560\ \text{cm}^{-1}$$

The respective values of D_e are thus

$$30{,}270 + \frac{1}{2}(2649.67)\ \text{cm}^{-1} = 31{,}590\ \text{cm}^{-1} \text{ for HBr}$$

$$89{,}560 + \frac{1}{2}(2170.21)\ \text{cm}^{-1} = 90{,}650\ \text{cm}^{-1} \text{ for CO}$$

Now to determine x_e and $v_e x_e$. For HBr, using equation 14.40:

$$x_e = \frac{2649.67\ \text{cm}^{-1}}{4(31{,}590\ \text{cm}^{-1})} = 0.02097 \qquad \therefore v_e x_e = (2649.67\ \text{cm}^{-1})(0.02097) = 55.6\ \text{cm}^{-1}$$

For CO, using equation 14.40:

$$x_e = \frac{2170.21\ \text{cm}^{-1}}{4(89{,}560\ \text{cm}^{-1})} = 0.006058 \qquad \therefore v_e x_e = (2170.21\ \text{cm}^{-1})(0.006058) = 13.2\ \text{cm}^{-1}$$

The predicted $v_e x_e$ value for HBr is off by about 20%, while that for CO is rather close.

14.47. Starting with equation 14.39, $E = hv_e\left(v + \dfrac{1}{2}\right) - hv_e x_e\left(v + \dfrac{1}{2}\right)^2$, we can construct a general expression for the change in energy between adjacent energy levels:

$$\Delta E = E(v+1) - E(v) = h\nu_e\left(v+\frac{3}{2}\right) - h\nu_e x_e\left(v+\frac{3}{2}\right)^2 - \left[h\nu_e\left(v+\frac{1}{2}\right) - h\nu_e x_e\left(v+\frac{1}{2}\right)^2\right]$$

Multiplying out the binomials and collecting terms (which won't be shown here), we get $\Delta E = h\nu_e - 2h\nu_e x_e(v+1)$. If we want to express this change of energy in frequency units directly (either in s^{-1} or cm^{-1}), we divide all terms by Planck's constant: $\nu = \nu_e - 2\nu_e x_e(v+1)$. For the fundamental absorption, $v = 0$ and this becomes $\nu = \nu_e - 2\nu_e x_e$. We know (see example 14.12) that we can use the ratio of reduced masses as a weighting factor to predict the shift of a vibrational frequency upon isotopic substitution. The frequency of the heavier isotope, ν_e*, is shifted by a factor $\sqrt{\dfrac{\mu}{\mu^*}}$. Defining this factor as ρ, we have that $\nu_e* = \rho\cdot\nu_e$. This variable also appears in the anharmonicity term as well. However, considering equation 14.40, we also find that the anharmonicity constant x_e is also defined in terms of ν_e, so x_e also reduces by the same factor of ρ. We can thus state that $x_e* = \rho\cdot x_e$. Assuming that, for the isotopically-substituted molecule, $\nu^* = \nu_e^* - 2\nu_e^* x_e^*$, we can substitute to get:

$\nu^* = \nu_e^* - 2\nu_e^* x_e^* = \rho\nu_e - 2(\rho\nu_e)(\rho x_e) = \rho\nu_e - 2\rho^2\nu_e x_e$, which is the expression from exercise 14.46.

14.49. The first vibration has σ_g^+ symmetry and is IR-inactive. The second vibration has σ_u^+ symmetry and is IR-active.

14.51. (a) E and C_3 are proper rotations; the σ_vs are improper rotations. (b) The E, C_3s, and C_2s are proper rotations; the S_4s and σ_ds are improper rotations. (c) The proper rotations are E, the C_6s, the C_3s, and the three classes of C_2s. The improper rotations are i, the S_3s, the S_6s, and all planes of symmetry. (d) The proper rotations are E and C_2, while the two S_4s are improper. (e) The proper rotations are E, the $C(\phi)$, and the infinite C_2s. The improper rotations are i, $S(\phi)$, and the infinite σ_vs. (f) All of the five classes of symmetry elements in O are considered proper rotations.

14.53. (a) H_2O_2 has C_2 symmetry. A check of the C_2 character table shows that both irreducible representations, A and B, will be IR-active. Therefore, H_2O_2 will have 6 IR-active vibrations. (b) Oxalic acid, $(COOH)_2$, has C_s symmetry. A check of the C_s character table shows that both irreducible representations, A' and A'', will be IR-active. Therefore, $(COOH)_2$ will have 18 IR-active vibrations. (c) Sulfur trioxide, SO_3, has D_{3h} symmetry, and not all of the irreducible representations are IR-active. Therefore, we must go through the steps to determine the character of the vibrations:

	E	$2C_3$	$3C_2$	σ_h	$2S_3$	$3\sigma_v$
$N_{stationary}$	4	1	2	4	1	2
θ	$0°$	$120°$	$180°$	$180°$	$60°$	$180°$
$1+2\cos\theta$	3	0	-1	-1	2	-1
$\pm N(1+2\cos\theta)$	12	0	-2	4	-2	2
$\chi_r, (1+2\cos\theta)$	3	0	-1	-1	2	-1
$\chi_{tr}, \pm(1+2\cos\theta)$	3	0	-1	1	-2	1
χ_{vib}	6	0	0	4	-2	2

Using the GOT, this reduces to $A_1'+2E'+A_2''$. According to the character table, E' and A_2'' modes are IR-active. Therefore, we expect SO_3 to have three IR-active vibrations.

(d) Formaldehyde is a C_{2v} molecule. Not all of the irreducible representations are IR-active. Therefore, we have to go through the steps to determine the characters of the vibrations:

	E	C_2	σ_v	$\sigma_v{}'$
$N_{stationary}$	4	2	4	2
θ	$0°$	$180°$	$180°$	$180°$
$1+2\cos\theta$	3	-1	-1	-1
$\pm N(1+2\cos\theta)$	12	-2	4	2
$\chi_r, (1+2\cos\theta)$	3	-1	-1	-1
$\chi_{tr}, \pm(1+2\cos\theta)$	3	-1	1	1
χ_{vib}	6	0	4	2

Using the GOT, this reduces to $3A_1+2B_1+B_2$, all of which are IR-active. Therefore, CH_2O will have 6 IR-active vibrations. (e) Acetone is a C_{2v} molecule. Not all of the irreducible representations are IR-active. Therefore, we have to go through the steps to determine the characters of the vibrations:

	E	C_2	σ_v	$\sigma_v{}'$
$N_{stationary}$	10	2	6	2
θ	$0°$	$180°$	$180°$	$180°$
$1+2\cos\theta$	3	-1	-1	-1
$\pm N(1+2\cos\theta)$	30	-2	6	2
$\chi_r, (1+2\cos\theta)$	3	-1	-1	-1
$\chi_{tr}, \pm(1+2\cos\theta)$	3	-1	1	1
χ_{vib}	24	0	6	2

Using the GOT, this reduces to $8A_1+4A_2+7B_1+5B_2$. Of these, the A_1, B_1, and B_2 vibrations are IR-active, so acetone will have 20 IR-active vibrations.

14.55. If KrF_4 were ever synthesized, one could tell if it were square planar or tetrahedral by determining the number of IR-active (and Raman-active) vibrations the molecule has. If the molecule were planar, it would have 3 IR-active vibrations, while if it were tetrahedral, it would have 4 IR-active vibrations.

14.57.

	E	$8C_3$	$3C_2$	$6C_4$	$6C_2'$	i	$8S_6$	$3\sigma_h$	$6S_4$	$6\sigma_d$
$N_{stationary}$	16	4	0	0	0	0	0	0	0	8
θ	0°	120°	180°	90°	180°	180°	120°	180°	90°	180°
$1+2\cos\theta$	3	0	-1	1	-1	-1	0	-1	1	-1
$\pm N(1+2\cos\theta)$	48	0	0	0	0	0	0	0	0	4
$\chi_r, (1+2\cos\theta)$	3	0	-1	1	-1	-1	0	-1	1	-1
$\chi_{tr}, \pm(1+2\cos\theta)$	3	0	-1	1	-1	1	0	1	-1	1
χ_{vib}	42	0	2	-2	2	0	0	0	0	8

Since only T_{1u} is IR-active, all we need to do is determine how many T_{1u}s there are in this reducible representation. Using the GOT, we find that there are $3T_{1u}$s as part of the irreducible representation. Therefore, there are three (triply-degenerate) IR-active vibrations. This represents 9 of the 42 possible vibrations of C_8H_8.

14.59. Combination bands are typically (but not always) weak because they are not usually allowed by the quantum-mechanical selection rules.

14.61. Both dioctyl sulfide and hexadecane are largely composed of long hydrocarbon chains, the only difference being that dioctyl sulfide has a sulfur atom in the middle. Thus, their vibrational spectra should be very similar.

14.63. SiH_4 is a spherical top whose rotational energy levels are given by the expression $\dfrac{J(J+1)\hbar^2}{2I}$ and whose spacing between rotational levels is $2B = \dfrac{2\hbar^2}{2I} = \dfrac{\hbar^2}{I}$. Since $2B = 16.72$ cm^{-1}, we convert this to joule units: $2B = 3.32\times10^{-22}$ J. Solving for I:

$$3.32\times10^{-22}\text{ J} = \frac{(6.626\times10^{-34}\text{ J}\cdot\text{s})^2}{4\pi^2 I} \qquad I = 3.35\times10^{-47}\text{ kg}\cdot\text{m}^2 = \sum m_i r_i^2$$

Now we need to determine a formula for the moment of inertia. If we assume that the molecule is spinning about an Si-H bond, then there are three equivalent H's making a triangle whose angle is determined by the tetrahedral bond angle, 109.45°. The angle that the bond makes from the rotational axis is half that, or 54.725°. The component of this bond that is rotating is related to the sine of this angle, or 0.8164. Finally, there are three hydrogens spinning. Thus, we have:

$$3.35\times10^{-47}\text{ kg}\cdot\text{m}^2 = 3(1.00\text{ g})\left(\frac{1\text{ kg}}{1000\text{ g}}\right)\left(\frac{1}{6.02\times10^{23}}\right)(0.8164r)^2 \quad \text{Solving for } r:$$

$r = 1.00\times10^{-10}$ m = 1.00 Å.

This information could not be determined from a pure rotational spectrum because SiH_4 does not have a permanent dipole moment, so does not have a pure rotational spectrum.

14.65. If centrifugal distortion were negligible, there would be no D_J terms in either equation 14.41 or 14.42. They would simplify to:

$$\Delta E = h\nu - 2x_e\nu_e + (B_1 + B_0)(J_{lower} + 1) + (B_1 - B_0)(J_{lower} + 1)^2 \qquad \text{and}$$

$$\Delta E = h\nu - 2x_e\nu_e - (B_1 + B_0)J_{lower} + (B_1 - B_0)J^2_{lower}$$

14.67. The spectra are different in that they will appear in different regions of the electromagnetic spectrum. The Raman spectrum generated using the He-Ne laser as excitation source will appear in the red part of the visible spectrum, while the Raman spectrum generated using the green light from the Kr^+ laser will appear in the green region of the visible spectrum. However, the pattern of new absorptions and their relative intensities should be virtually the same.

14.69. If a point group contains i, a center of inversion, then vibrations that are infrared-active are not Raman-active, and vibrations that are Raman-active are not infrared-active.

CHAPTER 15. INTRODUCTION TO ELECTRONIC SPECTROSCOPY AND STRUCTURE

15.1. Although you can check all of the irreducible representations in the D_{6h} character table individually, it is probably easiest to recognize that the only way the transition moment integral will be nonzero is if the allowed excited states also had E_{1u} symmetry. That's the only way that the product of the three irreducible representations will contain A_{1g}.

15.3. First, we need to determine the reduced masses of D and T. For D:

$$\mu = \frac{(9.109 \times 10^{-31} \text{ kg})(3.344 \times 10^{-27} \text{ kg})}{9.109 \times 10^{-31} \text{ kg} + 3.344 \times 10^{-27} \text{ kg}} = 9.1065 \times 10^{-31} \text{ kg}$$

For T: $\mu = \dfrac{(9.109 \times 10^{-31} \text{ kg})(5.008 \times 10^{-27} \text{ kg})}{9.109 \times 10^{-31} \text{ kg} + 5.008 \times 10^{-27} \text{ kg}} = 9.1073 \times 10^{-31} \text{ kg}$

Thus, the new Rydberg constants are:

$$R(\text{T}) = \frac{(1.602 \times 10^{-19} \text{ C})^4 (9.1065 \times 10^{-31} \text{ kg})}{8(8.854 \times 10^{-12} \text{ C}^4 / \text{J} \cdot \text{m})^2 (6.626 \times 10^{-34} \text{ J} \cdot \text{s})^2} = 2.1783 \times 10^{-18} \text{ J} = 109,660 \text{ cm}^{-1}$$

$$R(\text{T}) = \frac{(1.602 \times 10^{-19} \text{ C})^4 (9.1073 \times 10^{-31} \text{ kg})}{8(8.854 \times 10^{-12} \text{ C}^4 / \text{J} \cdot \text{m})^2 (6.626 \times 10^{-34} \text{ J} \cdot \text{s})^2} = 2.1785 \times 10^{-18} \text{ J} = 109,670 \text{ cm}^{-1}$$

These are very small – but detectable – changes.

15.5. The drawing is left to the student.

15.7. (a) The possible values of L are 2, 1, and 0. M_L goes from $-L$ to L and so can be at most –2, -1, 0, 1, and 2 (depending on the value of L). S can be 1 or 0, with M_S going from $-S$ to S (and so can be –1, 0, or 1, depending on S). J depends on the values of L and S, but can range from 3 to 0. M_J will depend on the value of J, but will range from $-J$ to J, so might be –3, -2, -1, 0, 1, 2, or 3. (b) The possible values of L are 6, 5, 4, 3, 2, 1, and 0. M_L goes from $-L$ to L and so can be at most –6 through 6 (depending on the value of L). S can be 1 or 0, with M_S going from $-S$ to S (and so can be –1, 0, or 1, depending on S). J depends on the values of L and S, but can range from 7 to 0. M_J will depend on the value of J, but will range from $-J$ to J, so might be –7 through 7. (c) The possible values of L are 3, 2, 1, and 0. M_L goes from $-L$ to L and so can be at most –3, -2, -1, 0, 1, 2, and 3 (depending on the value of L). S can be 1 or 0, with M_S going from $-S$ to S (and so can be –1, 0, or 1, depending on S). J depends on the values of L and S, but can range from 4 to 0. M_J will depend on the value of J, but will range from $-J$ to J, so might be –4, -3, -2, -1, 0, 1, 2, 3, or 4.

15.9. The aluminum atom has a single unpaired p electron in its valence shell. The spin of that single electron can show hyperfine coupling – that is, the spin can interact with the spin of the aluminum nucleus (which, by the way, has spin of 5/2) to exhibit closely-spaced but slightly different energy levels.

15.11. We need to convert the wavelengths into energy values and evaluate the difference:

$5890 \text{ A} \times \dfrac{1 \text{ m}}{10^{10} \text{ A}} = 5.890 \times 10^{-7} \text{ m} \qquad \nu = \dfrac{2.9979 \times 10^8 \text{ m/s}}{5.89 \times 10^{-7} \text{ m}} = 5.090 \times 10^{14} \text{ s}^{-1}$

84

$$E = (6.626\times10^{-34}\ \text{J}\cdot\text{s})(5.090\times10^{14}\ \text{s}^{-1}) = 3.373\times10^{-19}\ \text{J}$$

$$5896\ \text{A}\times\frac{1\,\text{m}}{10^{10}\ \text{A}} = 5.896\times10^{-7}\ \text{m} \qquad v = \frac{2.9979\times10^{8}\ \text{m/s}}{5.896\times10^{-7}\ \text{m}} = 5.085\times10^{14}\ \text{s}^{-1}$$

$$E = (6.626\times10^{-34}\ \text{J}\cdot\text{s})(5.085\times10^{14}\ \text{s}^{-1}) = 3.369\times10^{-19}\ \text{J}$$

The difference between these two energies is 4×10^{-22} J, which corresponds to about 240 J/mol.

15.13. According to a more mundane version of Hund's rules, electrons will spread out among available orbitals in a subshell before they pair in a single orbital, and will do so having the same orientation of spin. This "same orientation of spin" is what yields a maximum multiplicity of the ground state. In such a situation, however, the overall L – as determined by the vector sum of all of the m_l values – is zero, as they all cancel. Thus, half-filled subshells will always have their highest multiplicity for an S term symbol.

15.15. An h electron has $\ell = 5$; therefore, possible L values for two h electrons are 0, 1, 2, 3, 4, 5, 6, 7, 8, 9, and 10. These lead to S, P, D, F, G, H, I, J, K, and L terms. For two electrons, the S values can be either 0 or 1, leading to the combinations ^{3}S, ^{1}S, ^{3}P, ^{1}P, ^{3}D, ^{1}D, ^{3}F, ^{1}F, ^{3}G, ^{1}G, ^{3}H, ^{1}H, ^{3}I, ^{1}I, ^{3}J, ^{1}J, ^{3}K, ^{1}K, ^{3}L, and ^{1}L. If we were to include J values, the term symbols would be ^{3}S$_{0,1}$, ^{1}S$_0$, ^{3}P$_{2,1,0}$, ^{1}P$_1$, ^{3}D$_{3,2,1}$, ^{1}D$_2$, ^{3}F$_{4,3,2}$, ^{1}F$_3$, ^{3}G$_{5,4,3}$, ^{1}G$_4$, ^{3}H$_{6,5,4}$, ^{1}H$_5$, ^{3}I$_{7,6,5}$, ^{1}I$_6$, ^{3}J$_{8,7,6}$, ^{1}J$_7$, ^{3}K$_{9,8,7}$, ^{1}K$_8$, ^{3}L$_{10,9,8}$, and ^{1}L$_9$. (Only one J value would be present for each individual term; the J values for the triplet states are grouped together for reasons of clarity.) We are not excluding any possible term symbol, but some of the above may be forbidden by Pauli principle arguments. According to Hund's rules, the higher multiplicity is preferred for the ground state (3, here); the higher value of L is preferred (L, here), and for a less-than-half-filled subshell, the lower value of J is preferred (8, here). So the ground state term symbol should be ^{3}L$_8$.

15.17. Since $\Delta S = 0$, any allowed excited state will have $S = 1$. $\Delta L = 0$ or ± 1, so allowed excited states can be either D, F, or G terms. Finally, since $\Delta J = 0$ or ± 1, allowed excited states can have a value for J of 1, 2, or 3. Combining these three items, the possible term symbols are ^{3}D$_1$, ^{3}D$_2$, ^{3}D$_3$, ^{3}F$_1$, ^{3}F$_2$, ^{3}F$_3$, ^{3}G$_1$, ^{3}G$_2$, or ^{3}G$_3$. Some of these aren't possible due to the relationship between L, S, and J: only ^{3}D$_1$, ^{3}D$_2$, ^{3}D$_3$, ^{3}F$_2$, ^{3}F$_3$, or ^{3}G$_3$ are valid term symbols.

15.19. A diatomic molecule is, in a first approximation, a rigid rotor. Thus, its behavior can be modeled to some degree of accuracy by the 3-dimensional rigid rotor ideal system, which ultimately predicts that the angular momentum of the system is quantized.

15.21. If Li$_2$ has two electrons in π_u orbitals, then the term symbols derive from the product $\Pi_u \times \Pi_u$ in the D$_{\infty h}$ point group. This gives us the set of characters $(4, 4\cos^2\phi, 0, 4, 4\cos^2\phi, 0)$. By analogy to the discussion of the term symbols of diatomic molecules in the text, we rationalize the irreducible representations that contribute to this character set, rather than use the GOT directly. Using the same arguments, we should get $\Sigma_g^+ + \Sigma_g^- + \Delta_g$.

15.23. The acetylide ion, C$_2^{2-}$, is isoelectronic with O$_2$, so it should have the same molecular orbital diagram. Thus, it should have the same ground-state term symbol, $^3\Sigma_g^-$.

15.25. (a) No unpaired electrons, so NO_3^- should be colorless. (b) MnO_4^- has unpaired electrons in the Mn's d subshell, so it is highly likely that the ion will absorb visible light and have some color. (It is, in fact, strongly purple.) (c) There are no unpaired electrons, so NH_4^+ should be colorless. (d) $Cr_2O_7^{2-}$ has unpaired electrons in Cr's d subshell, so it is highly likely that the ion will absorb visible light and have some color. (It is, in fact, yellow-orange.) (e) There are no unpaired electrons in O_2^{2-}, so we expect it to be colorless. (f) Acetylide ions have unpaired electrons in its molecular orbitals (see the answer to exercise 15.23), so we expect it might have a color. (It's grayish-black in color.)

15.27. According to the point group character table in the Appendix, the electric dipole operator has the irreducible representations A_1, B_1, or B_2 (depending on which three-dimensional axis is involved). In order to guarantee that the A_1 irreducible representation is part of the product $\Psi_{upper} \hat{\mu} \Psi_{lower}$, an excited state must also have the irreducible representations A_1, B_1, or B_2. Thus, possible excited states will have irreducible representations $^1A_1$, $^1B_1$, or $^1B_2$.

15.29. Nothing would change in the Hückel approximation of ethylene if deuterium atoms were substituted for hydrogen atoms in the molecule. Deuterium has the same electronic structure as hydrogen (with the exception of a small change in the reduced mass of the system), so substitution shouldn't change the electronic approximation.

15.31. Cyclopentadiene easily accepts an electron because when it does, it has six π electrons and is aromatic.

15.33. (a) Ideally, neutral cyclopolyenes will be aromatic if they have $4n+2$ carbon atoms with π electrons in the ring. (b) Ideally, cyclopolyenes having a single positive charge if they have $4n+3$ carbon atoms with π electrons in the ring. (c) Ideally, cyclopolyenes having a double positive charge if they have $4n+4$ carbon atoms with π electrons in the ring.

15.35. Heating a potentially laser-active substance excites systems (atoms and molecules) into excited states in a manner that ultimately mimics thermal equilibrium. Thermal equilibrium is not a population inversion; the higher an energy level is, the less populated it is. Thus, it would be extremely difficult (if not impossible) to construct a population inversion by purely thermal means; some other method would be necessary.

15.37. Light from fireflies would be a (chemically-induced) example of phosphorescence.

15.39. $10.6\ \mu m \times \dfrac{1\ m}{10^6\ \mu m} = 1.06 \times 10^{-5}\ m$ $\qquad v = \dfrac{c}{\lambda} = \dfrac{2.9979 \times 10^8\ m/s}{1.06 \times 10^{-5}\ m} = 2.83 \times 10^{13}\ s^{-1}$

$E = h\nu = (6.626 \times 10^{-34}\ J \cdot s)(2.83 \times 10^{13}\ s^{-1}) = 1.87 \times 10^{-20}\ J$ per photon

Therefore, $\dfrac{300,000\ J}{s} \times \dfrac{1\ photon}{1.87 \times 10^{-20}\ J} = 1.60 \times 10^{25}$ photons per second.

15.41. Power is defined as energy per unit time, so for a 300 mJ pulse in 2.50 ns:

$2.50\ ns \times \dfrac{1\ s}{10^9\ ns} = 2.50 \times 10^{-9}\ s$ $\qquad 300\ mJ \times \dfrac{1\ J}{1000\ mJ} = 3.00 \times 10^{-1}\ J$

$$\text{power} = \frac{3.00 \times 10^{-1} \text{ J}}{2.50 \times 10^{-9} \text{ s}} = 1.20 \times 10^{8} \text{ J/s} = 120 \text{ MW (megawatts)}$$

CHAPTER 16. INTRODUCTION TO MAGNETIC SPECTROSCOPY

16.1. A magnetic field vector is the magnetic phenomenon formed when a current I flows through a straight wire, forming a circular magnetic field. A magnetic dipole vector is a linear magnetic effect formed by a current I flowing in a circle.

16.3. (a) If the electron had a positive charge, the value of the magnetic dipole wouldn't change because its value is based on the magnitude of the charge, not its positivity or negativity. However, because the charge is opposite, the direction of the magnetic dipole vector will be opposite its original direction. (b) For positronium, we can argue that the magnetic dipole will be zero overall. That's because in positronium, we have two particles "orbiting" each other that have the same mass, and rather than having the tiny electron moving around a massive, unmoving nucleus, we have two equal-mass particles in orbit about a mutual center. The orbits will be, on average, equal – producing equal magnetic dipoles – but opposite – implying that the two magnetic dipole vectors will cancel each other out.

16.5. According to equation 16.6, the Bohr magneton μ_B equals $\dfrac{e\hbar}{2m_e}$. Substituting for these

constants: $\mu_B = \dfrac{(1.602\times10^{-19}\ \text{C})(6.626\times10^{-34}\ \text{J}\cdot\text{s})}{2\pi\cdot2(9.109\times10^{-31}\ \text{kg})} = 9.273\times10^{-24}\ \text{C}\cdot\text{J}\cdot\text{s/kg}$

Within truncation error, this is the same value as given in the text. However, the units are unusual. But if we recall that a tesla is defined as a kg/C·s, we note that the units of μ_B are equivalent to J/T. Thus, we verify both the value and units of the Bohr magneton.

16.7. (a) Of the three states in the $^1S \rightarrow {}^1P$ transition, one state decreases in energy, one state increases in energy by the same amount, and one state doesn't change in energy at all. The change in energy is given by equation 16.9: $\Delta E = \mu_B\cdot M_L\cdot B$:
$\Delta E = (9.274\times10^{-24}\ \text{J/T})(\pm1)(2.35\ \text{T}) = \pm2.18\times10^{-23}\ \text{J}$. This corresponds to 1.10 cm^{-1}.
(b) For the $^1P \rightarrow {}^1D$ transition in the presence of a magnetic field, there will be 15 possible transitions, as the ^1P state as three possible M_L values while the ^1D state has 5 possible M_L values. However, many of the energy differences will be the same, and the selection rules in equation 16.8 will limit the number of allowed transitions. In the end, there will be only three lines having, not surprisingly the same ΔE as for the $^1S \rightarrow {}^1P$ transition: $\Delta E = \pm2.18\times10^{-23}\ \text{J}$.

16.9. We need the g_J value for the $^2P_{3/2}$ term symbol. Using equation 16.13:
$g_J = 1 + \dfrac{(3/2)(3/2+1)+(1/2)(1/2+1)-(1)(1+1)}{2(3/2)(3/2+1)} = 1.3333...$
Now we use equation 16.11 and the fact that the maximum M_J value is $+3/2$ and the minimum is $-3/2$ to calculate the energy difference between these two states (the difference being equal to 3):
$\Delta E = (1.3333)(9.274\times10^{-24}\ \text{J/T})(3)(0.00006\ \text{T}) = 2.23\times10^{-27}\ \text{J}$. This is equivalent to 1.12×10^{-4} cm^{-1}, a difficult energy difference to differentiate.

16.11. Using equation 16.12:

$$g_J = 1 + \left[\frac{(4)(4+1) + (2)(2+1) - (2)(2+1)}{2(4)(4+1)} \right](2.0023 - 1) = 1.50115$$

Using equation 16.13:

$$g_J = 1 + \left[\frac{(4)(4+1) + (2)(2+1) - (2)(2+1)}{2(4)(4+1)} \right] = 1.5$$

The difference is less than 0.08%.

16.13. In several places, the text suggests that the frequency of microwave radiation used in ESR spectroscopy is on the order of 10 GHz. Let us use that frequency to determine an approximate energy per photon:

$E = (6.626 \times 10^{-34} \text{ J})(10 \times 10^9 \text{ s}^{-1}) = 7 \times 10^{-24}$ J/photon.

Of course, the energy depends on the frequency of microwave radiation used. This can be compared to $3 \times 10^{-19} - 5 \times 10^{-19}$ J for a single photon of visible light.

16.15. For the amine radical ($NH_2 \cdot$), we can consider Example 16.7b of $NH_3 \cdot$, the ammonia radical. In this example, we had a maximum possible M_I of 5/2, leading to $2(5/2) + 1 = 6$ ESR signals. For the amine radical, we have one less hydrogen, so a maximum possible M_I of 2, leading to $2(2) + 1 = 5$ ESR signals.

16.17. Since the carbon atoms have $I = 0$, they do not contribute to the hyperfine coupling of the electron. However, the seven hydrogen atoms all have $I = \frac{1}{2}$, and the possible combinations of those spins can lead to M_I values of +7/2, +5/2, +3/2, +1/2, -1/2, -3/2, -5/2, and –7/2. Thus, we would expect 8 ESR signals for the cycloheptatrienyl radical.

16.19. $\Delta E = (2.0058)(9.274 \times 10^{-24} \text{ J/T})(0.3476 \text{ T})$ $\Delta E = 6.466 \times 10^{-24}$ J

Converting this to wavenumbers, we find that $\Delta E = 0.3255$ cm^{-1}.

16.21. With six lines in the ESR spectrum, an atom in the radical can have a spin of up to 5/2 (assuming that there are other atoms in the radical that are contributing to the signal); if it has a spin higher than that, then that particular atom probably isn't in the molecule. Checking the table of nuclear spins, we find that (a) ^{42}K ($I = 2$), (b) ^{35}Cl ($I = 3/2$), (c) ^{37}Cl ($I = 3/2$), (d) ^{67}Zn ($I = 5/2$), (e) ^{47}Ti ($I = 5/2$), and (f) ^{32}S ($I = 0$) are all possible in the radical. In the cases of ^{67}Zn and ^{47}Ti, no other atoms in the radical should have a non-zero nuclear spin; in all other cases, other atoms with non-zero I are necessary in order to produce the given ESR spectrum.

16.23. The NMR transitions should appear at the same wavelength. Equations 16.22 and 16.23, which give expressions for the frequency of resonance, do not have a dependence on the M_I values (since $\Delta M_I = \pm 1$).

16.25. Since the working frequency is given in MHz, we will use the hertz version of the resonance condition, equation 16.23. Nuclear g_N values are found in Table 16.1 and the appendix:

For ^{2}H: $330 \times 10^6 \text{ s}^{-1} = \dfrac{(0.85744)(5.051 \times 10^{-27} \text{ J/T})B}{6.626 \times 10^{-34} \text{ J} \cdot \text{s}}$ $B = 50.49$ T

For ^{19}F: $330 \times 10^6 \text{ s}^{-1} = \dfrac{(5.2567)(5.051 \times 10^{-27} \text{ J/T})B}{6.626 \times 10^{-34} \text{ J} \cdot \text{s}}$ $B = 8.235 \text{ T}$

For ^{31}P: $330 \times 10^6 \text{ s}^{-1} = \dfrac{(2.2634)(5.051 \times 10^{-27} \text{ J/T})B}{6.626 \times 10^{-34} \text{ J} \cdot \text{s}}$ $B = 19.13 \text{ T}$

For ^{55}Mn: $330 \times 10^6 \text{ s}^{-1} = \dfrac{(1.387)(5.051 \times 10^{-27} \text{ J/T})B}{6.626 \times 10^{-34} \text{ J} \cdot \text{s}}$ $B = 31.20 \text{ T}$

16.27. (a) An acceptable molecule that yields this NMR spectrum is CH_3COOCH_3, or methyl acetate. (b) An acceptable molecule that yields this NMR spectrum is 1-bromo-2-methylpropane, or $CH_3CH(CH_3)CH_2Br$.

16.29. The differences between energy levels is the same, since there is a unit change in the M_I quantum number for all of the allowed transitions. Using equation 16.23:

$$\nu = \frac{(0.5479)(5.051 \times 10^{-27} \text{ J/T})(3.45 \text{ T})}{6.626 \times 10^{-34} \text{ J} \cdot \text{s}} = 1.441 \times 10^7 \text{ s}^{-1} = 14.41 \text{ MHz}$$

16.31. $2.45 \times 10^9 \text{ s}^{-1} = \dfrac{(2.0823)(5.051 \times 10^{-27} \text{ J/T})B}{6.626 \times 10^{-34} \text{ J} \cdot \text{s}}$ $B = 154.35 \text{ T}$

CHAPTER 17. STATISTICAL MECHANICS: INTRODUCTION

17.1. There are four ways of putting one of three balls in each of four boxes. It doesn't agree with equation 17.1 because we are assuming that the balls are indistinguishable. If they are distinguishable, then there are 24 different ways, which is consistent with equation 17.1. If there are no restrictions on the number of balls in each box, then there are 20 different ways of putting the balls in the boxes.

17.3. According to Stirling's approximation, the natural logarithm of 1,000,000! is
$\ln(1{,}000{,}000!) = (1{,}000{,}000)\ln(1{,}000{,}000) - 1{,}000{,}000 = 12{,}816{,}000$. Therefore, 1,000,000! is about $e^{12{,}816{,}000}$, or about $10^{5{,}566{,}000}$.

17.5. The logarithm of the expression is
$$\ln(N-1)! = \ln\left[e^{-N} \cdot N^{N-1/2} \cdot (2\pi)^{1/2} \cdot \left(1 + \frac{1}{12N} + \frac{1}{288N^2} + \cdots \right) \right] \quad \text{Using properties of logarithms,}$$

this reduces to
$$\ln(N-1)! = -N + \left(N - \frac{1}{2} \right) \ln N + \frac{1}{2} \ln 2\pi + \ln\left(1 + \frac{1}{12N} + \frac{1}{288N^2} \right). \quad \text{To evaluate } \ln(5000!), \text{ we set}$$

N to 5001 and substitute:
$$\ln(5001-1)! = -5001 + \left(5001 - \frac{1}{2} \right) \ln 5001 + \frac{1}{2} \ln 2\pi + \ln\left(1 + \frac{1}{12(5001)} + \frac{1}{288(5001)^2} \right)$$

Evaluating each term:
$\ln(5000)! = -5001 + 42{,}591.22 + 0.9189 + 1.6663 \times 10^{-5} = 37{,}591$ (This is one of the values that's listed in the table in chapter 17, so this expression is much more accurate than Stirling's approximation.)

17.7. According to the description in the exercise, we need to solve the following expression:
$$\frac{\int_0^{12} \left[-(7-x)^2 + 38 \right] dx}{12} \quad \text{Expanding and evaluating:}$$

$$\frac{\int_0^{12} \left[-49 + 14x - x^2 + 38 \right] dx}{12} = \frac{\int_0^{12} \left(14x - x^2 - 11 \right) dx}{12} = \frac{\left(7x^2 - \frac{1}{3}x^3 - 11x \right)\Big|_0^{12}}{12}$$

$$= \frac{1}{12}\left(7 \cdot 144 - \frac{1728}{3} - 11 \cdot 12 - 0 \right) = 25. \quad \text{Thus, on average, there will be 25 insects per month.}$$

17.9. For a grand canonical ensemble, equations 17.4 – 17.6 would be rewritten as
$$V = \sum_j V_j = j \cdot V_j \qquad T = T_j \qquad \mu = \mu_j$$

The last relationship is due to the fact that chemical potential is also an intensive property.

17.11. If we refer to Figure 17.6, we see that the distribution labeled (1,0,1,1) appears six times out of twelve, making it the most probable distribution (with a probability of 0.50). For this system, the thermodynamic properties are probably not dictated by this one distribution, because there are only twelve possible distributions and the other distributions will have some obvious effect on the overall average. But for a system that has a larger number of possible distributions, the most probable distribution has a larger and larger impact on the overall thermodynamic properties of the system as a whole.

17.13. (a) $\phi = C_1\xi_1 + C_2\xi_2 + C_3\xi_3$ is the complete equation. Taking the derivative with respect to ξ_1: $\dfrac{\partial\phi}{\partial\xi_1} = \dfrac{\partial(C_1\xi_1)}{\partial\xi_1} + \dfrac{\partial(C_2\xi_2)}{\partial\xi_1} + \dfrac{\partial(C_3\xi_3)}{\partial\xi_1} = C_1 + 0 + 0 = C_1$. Similar expressions exist for

derivatives with respect to ξ_2 and ξ_3. (b) The general expression is $\dfrac{\partial\phi}{\partial\xi_i} = C_i$. This general

expression was applied three times, once for each summation in equation 17.14, to get the four remaining terms (in one, the chain rule of derivation was applied) that resulted in equation 17.15.

17.15. Use equation 17.21: $\dfrac{N_1}{N_0} = \dfrac{g_1}{g_0}e^{-\Delta E/kT}$. If we assume the degeneracies are the same, the

g_1/g_0 term cancels. Converting 200 cm^{-1} to J units:

$\lambda = \dfrac{1}{\tilde{\nu}} = \dfrac{1}{200\,\text{cm}^{-1}} = 0.005\,\text{cm} = 0.00005\ \text{m} \qquad \nu = \dfrac{2.9979\times10^8\ \text{m/s}}{0.00005\,\text{m}} = 5.996\times10^{12}\ \text{s}^{-1}$

$E = h\nu = (6.626\times10^{-34}\ \text{J·s})(5.996\times10^{12}\ \text{s}^{-1}) = 3.973\times10^{-21}\ \text{J}$

Now we can use equation 17.21: $\dfrac{N_1}{N_0} = e^{-(3.973\times10^{-21}\ \text{J})/(1.381\times10^{-23}\ \text{J/K})(298\ \text{K})} = 0.38$.

Thus, there are almost three times as many atoms in the ground state as there are in the first excited state.

17.17. By various conversions (see exercise 17.15 above), we can show that 16.4 cm^{-1} = 3.26×10^{-22} J and 43.5 cm^{-1} = 8.64×10^{-22} J. Using equation 17.21, to have twice as many atoms in the ground state as the first electronic state:

$\dfrac{1}{2} = \dfrac{3}{1}e^{-(3.26\times10^{-22}\ \text{J})/(1.381\times10^{-23}\ \text{J/K})(T)}$ Solving for T (which includes taking the natural logarithm of

both sides): $-1.79175\ldots = -\dfrac{3.26\times10^{-22}}{(1.381\times10^{-23})T} \qquad T = 13.2\ \text{K}$

To have equal populations in the ground state and second excited state:

$1 = \dfrac{5}{1}e^{-(8.64\times10^{-22}\ \text{J})/(1.381\times10^{-23}\ \text{J/K})(T)}$ Solving for T: $-1.60943\ldots = -\dfrac{8.64\times10^{-22}}{(1.381\times10^{-23})T} \qquad T = 38.9\ \text{K}$

In order to determine the temperature for equal populations in the first and second excited states, we need the energy difference between the two: 8.64×10^{-22} J − 3.26×10^{-22} J = 5.38×10^{-22} J. Solving:

$$1 = \frac{5}{3} e^{-(5.38 \times 10^{-22} \text{ J})/(1.381 \times 10^{-23} \text{ J/K})(T)}$$ Solving for T: $-0.51082\ldots = -\dfrac{8.64 \times 10^{-22}}{(1.381 \times 10^{-23})T}$ $T = 122.5 \text{ K}$

17.19. (a) The state function A will always have a slightly lower value (i.e. a larger magnitude negative value) than G, based on the definitions in equations 17.44 and 17.45. (b) It is difficult to tell which is greater because the expressions are so different. One can imagine circumstances in which either is greater than the other.

17.21. For equation 17.46, $\mu_i = -kT \ln \dfrac{q}{N_i}$, the restriction is that the amount of any other substance in the system, n_j ($j \neq i$), must remain constant. This is the same restriction we placed on the chemical potential when it was introduced earlier in the text.

17.23. (a) From a statistical perspective, we can argue that there are more energy states available when a gas expands, so q would be larger, meaning that S would increase. (b) At 5°C, there is enough energy available to allow the molecules to access a larger number of energy states, suggesting an increase in q and a concurrent increase in S.

17.25. We need to show that $\displaystyle\lim_{T \to 0} \left[k \ln\left(\sum g_i e^{-\varepsilon_i/kT}\right) + \frac{1}{T} \frac{\sum \varepsilon_i g_i e^{-\varepsilon_i/kT}}{\sum g_i e^{-\varepsilon_i/kT}} \right] = k \ln g_0$. The first term

is easy. Consider the summation. The first term in the summation is g_0, because ε_0 in the exponent is zero (as the ground state) and the limit of that term as T goes to zero is just g_0. For all of the remaining terms, as T goes to zero, the exponentials become $e^{-\infty}$, which is zero. So for the first term, we get $k \ln g_0$ as T goes to zero. The second term is more problematic, and we will need to apply L'Hopital's rule. Let us rewrite the second term by factoring out the exponential with respect to temperature out of every term in the summation:

$$\frac{1}{T} \frac{\sum \varepsilon_i g_i e^{-\varepsilon_i/kT}}{\sum g_i e^{-\varepsilon_i/kT}} = \frac{1}{T} \frac{e^{-1/kT} \sum \varepsilon_i g_i e^{\varepsilon_i}}{e^{-1/kT} \sum g_i e^{\varepsilon_i}} = \frac{e^{-1/kT} \sum \varepsilon_i g_i e^{\varepsilon_i}}{T e^{-1/kT} \sum g_i e^{\varepsilon_i}}$$

By factoring the $e^{-1/kT}$ out of each term, we are taking advantage of a property of exponentials. All we need to do now is determine the limit of $\dfrac{e^{-1/kT}}{T e^{-1/kT}}$ as T approaches 0. We can't simply substitute $T = 0$ into the expression because we will have trouble with infinities and zeros. However, we can use L'Hopital's rule and take the derivative of the numerator and denominator:

$$\frac{D(e^{-1/kT})}{D(T e^{-1/kT})} = \frac{\dfrac{1}{kT^2} e^{-1/kT}}{e^{-1/kT} + T \dfrac{1}{kT^2} e^{-1/kT}} = \frac{\dfrac{1}{kT^2}}{1 + \dfrac{1}{kT}} = \frac{\dfrac{1}{kT^2}}{\dfrac{kT+1}{kT}} = \frac{1}{kT^2} \cdot \frac{kT}{kT+1} = \frac{1}{kT^2 + T}$$. Evaluating this

expression at the limit of $T = 0$ is still problematic, but we can apply L'Hopital's rule again to get

93

$$\frac{D(1)}{D(kT^2 + T)} = \frac{0}{2kT + 1},$$ which is an expression that can be evaluated at $T = 0$. The fraction equals 0 at $T = 0$, so the entire second term goes to zero as T goes to zero, and we are left with the fact that $\lim_{T \to 0} S = k \ln g_0$.

17.27. If $N = N_A$ in the Sackur-Tetrode equation, the equation itself doesn't change, but the value of S calculated using that value is now the molar entropy.

17.29. The derivation of equation 17.56 starts from equation 17.34: $E = NkT^2 \left(\frac{1}{q} \frac{\partial q}{\partial T} \right)$. First, let us evaluate the derivative of q with respect to T:

$$\frac{\partial q}{\partial T} = \frac{\partial}{\partial T} \left[\left(\frac{2\pi mkT}{h^2} \right)^{3/2} V \right] = \frac{3}{2} \left(\frac{2\pi mk}{h^2} \right)^{3/2} VT^{1/2} \qquad \text{Now, dividing this by } q:$$

$$\frac{\frac{3}{2} \left(\frac{2\pi mk}{h^2} \right)^{3/2} VT^{1/2}}{\left(\frac{2\pi mkT}{h^2} \right)^{3/2} V}.$$ Complex as this looks, everything cancels except $\frac{3}{2T}$! Substituting back

into the original equation: $E = NkT^2 \cdot \frac{3}{2T} = \frac{3}{2} NkT$, which is equation 17.56.

17.31. At 25 K: $\Lambda = \left(\frac{(6.626 \times 10^{-34} \text{ J} \cdot \text{s})^2}{2\pi \left(\frac{0.004 \text{ kg}}{6.02 \times 10^{23}} \right) (1.381 \times 10^{-23} \text{ J/K})(25 \text{ K})} \right)^{1/2} = 1.745 \times 10^{-10} \text{ m}$

At 500 K: $\Lambda = \left(\frac{(6.626 \times 10^{-34} \text{ J} \cdot \text{s})^2}{2\pi \left(\frac{0.004 \text{ kg}}{6.02 \times 10^{23}} \right) (1.381 \times 10^{-23} \text{ J/K})(500 \text{ K})} \right)^{1/2} = 3.903 \times 10^{-11} \text{ m}$

17.33. (a)

$$S = (6.02 \times 10^{23})(1.381 \times 10^{-23} \text{ J/K}) \left[\ln \left\{ \left(\frac{2\pi(0.012 \text{ kg}/6.02 \times 10^{23})(1.381 \times 10^{-23} \text{ J/K})(1000 \text{ K})}{(6.626 \times 10^{-34} \text{ J} \cdot \text{s})^2} \right)^{3/2} \right. \right.$$

$$\times \frac{(1.381 \times 10^{-23} \text{ J/K})(1000 \text{ K})}{(1.000 \text{ atm})} \cdot \frac{1 \text{ L} \cdot \text{atm}}{101.32 \text{ J}} \cdot \frac{1 \text{ m}^3}{1000 \text{ L}} \right\} + \frac{5}{2} \right]$$

Solving: $S = 164.9$ J/K   J mol^{-1} k^{-1}

94

(b) $S = (6.02 \times 10^{23})(1.381 \times 10^{-23} \text{ J/K}) \left[\ln \left\{ \left(\dfrac{2\pi(0.0558 \text{ kg}/6.02 \times 10^{23})(1.381 \times 10^{-23} \text{ J/K})(3500 \text{ K})}{(6.626 \times 10^{-34} \text{ J} \cdot \text{s})^2} \right)^{3/2} \right. \right.$

$\left. \left. \times \dfrac{(1.381 \times 10^{-23} \text{ J/K})(3500 \text{ K})}{(1.000 \text{ atm})} \cdot \dfrac{1 \text{ L} \cdot \text{atm}}{101.32 \text{ J}} \cdot \dfrac{1 \text{ m}^3}{1000 \text{ L}} \right\} + \dfrac{5}{2} \right]$

$S = 210.1$ J/K

(c) $S = (6.02 \times 10^{23})(1.381 \times 10^{-23} \text{ J/K}) \left[\ln \left\{ \left(\dfrac{2\pi(0.2006 \text{ kg}/6.02 \times 10^{23})(1.381 \times 10^{-23} \text{ J/K})(298 \text{ K})}{(6.626 \times 10^{-34} \text{ J} \cdot \text{s})^2} \right)^{3/2} \right. \right.$

$\left. \left. \times \dfrac{(1.381 \times 10^{-23} \text{ J/K})(298 \text{ K})}{(1.000 \text{ atm})} \cdot \dfrac{1 \text{ L} \cdot \text{atm}}{101.32 \text{ J}} \cdot \dfrac{1 \text{ m}^3}{1000 \text{ L}} \right\} + \dfrac{5}{2} \right]$

$S = 174.9$ J/K.

17.35. First, we can calculate the de Broglie wavelength of electrons going $0.01c$ ($= 2.9979 \times 10^6$ m/s): $\lambda = \dfrac{6.626 \times 10^{-34} \text{ J} \cdot \text{s}}{(9.109 \times 10^{-31} \text{ kg})(2.9979 \times 10^6 \text{ m/s})} = 2.43 \times 10^{-10}$ m

Next, we use this as the wavelength in the thermal de Broglie equation and solve for T:

$2.43 \times 10^{-10} \text{ m} = \left(\dfrac{(6.626 \times 10^{-34} \text{ J} \cdot \text{s})^2}{2\pi \cdot (9.109 \times 10^{-31} \text{ kg})(1.381 \times 10^{-23} \text{ J/K})T} \right)^{1/2}$

$T = 94,100$ K

CHAPTER 18. MORE STATISTICAL MECHANICS

18.1. (a) ^{12}C has a nuclear spin of zero, so there are $2(0) + 1 = 1$ nuclear spin state. Therefore, $q_{nuc} = 1$. (b) ^{56}Fe has a nuclear spin of zero, so there are $2(0) + 1 = 1$ nuclear spin state. Therefore, $q_{nuc} = 1$. (c) ^{1}H has a nuclear spin of ½ , so there are $2(½) + 1 = 2$ nuclear spin states. Therefore, $q_{nuc} = 2$. (d) ^{2}H has a nuclear spin of 1, so there are $2(1) + 1 = 3$ nuclear spin states. Therefore, $q_{nuc} = 3$.

18.3. For both E and S, the nuclear contribution is zero. That's because both E and S are determined as a derivative of the partition function. In our approximation, q_{nuc} is a constant (equal to the degeneracy of the ground nuclear state), so the derivative of this constant is simply zero.

18.5. The minimum value of q_{elect} is 1, because this is the minimum degeneracy of any electronic ground state.

18.7. According to example 18.3, the value of D_0 is
$$\frac{432\,kJ}{mol} \cdot \frac{1000\,J}{kJ} \cdot \frac{1\,mol}{6.02 \times 10^{23}\,molecules} = 7.18 \times 10^{-19}\,J$$
for one molecule of H_2. Using this value in the definition of the partition function:
$$q_{elect} = 1 \cdot e^{7.18 \times 10^{-19}\,J/(1.381 \times 10^{-23}\,J/K)(298\,K)} = 5.89 \times 10^{75}.$$ This is somewhat lower than the original value of about 2×10^{80}, but still a large number.

18.9. (a) Assuming that the 89.8 J is the value for D_e (to be accurate, we would need to include the zero-point vibrational energy):
$$\frac{89.8\,J}{mol} \cdot \frac{1\,mol}{6.02 \times 10^{23}\,molecules} = 1.49 \times 10^{-22}\,J$$ Substituting into q_{elect}:
$$q_{elect} = 1 \cdot e^{1.49 \times 10^{-22}\,J/(1.381 \times 10^{-23}\,J/K)(4.2\,K)} = 13.1$$
(b) Room temperature is enough thermal energy ($= RT = 2.48$ kJ/mol) to break such a weak bond, so He_2 probably won't exist at such "high" temperatures.

18.11. The ratio of $q_{vib}(H_2)$ to $q_{vib}(D_2)$ can be written using equation 18.20:
$$\frac{q_{vib}(H_2)}{q_{vib}(D_2)} = \frac{(kT/h\nu_{H_2})}{(kT/h\nu_{D_2})} = \frac{\nu_{D_2}}{\nu_{H_2}}.$$ Since $\nu = \frac{1}{2\pi}\sqrt{\frac{k}{\mu}}$, where k is the force constant and μ is the

reduced mass, this ratio of partition functions becomes $\dfrac{q_{vib}(H_2)}{q_{vib}(D_2)} = \dfrac{(1/2\pi)\sqrt{k/\mu_{D_2}}}{(1/2\pi)\sqrt{k/\mu_{H_2}}}$, which

reduces to $\dfrac{q_{vib}(H_2)}{q_{vib}(D_2)} = \sqrt{\dfrac{\mu_{H_2}}{\mu_{D_2}}}$. We can substitute the reduced masses for H_2 and D_2 into this

expression or, recalling that deuterium has twice the mass of hydrogen, recognize that deuterium will have twice the reduced mass of hydrogen. Thus, the ratio simplifies and we get our final answer: $\dfrac{q_{vib}(H_2)}{q_{vib}(D_2)} = \sqrt{\dfrac{1}{2}}$.

18.13. $q_{vib} = \prod_1^{3N-6}\left(\dfrac{e^{-\theta/2T}}{1-e^{-\theta/T}}\right) = \left(\dfrac{e^{-1870/2\cdot298}}{1-e^{-1870/298}}\right)^3\left(\dfrac{e^{-2180/2\cdot298}}{1-e^{-2180/298}}\right)^2\left(\dfrac{e^{-4170/2\cdot298}}{1-e^{-4170/298}}\right)\left(\dfrac{e^{-4320/2\cdot298}}{1-e^{-4320/298}}\right)^3$

$q_{vib} = (0.04347)^3(0.02581)^2(0.0009149)(0.0007114)^3 = 1.802\times10^{-20}$

18.15. For a gas-phase molecule, the minimum value of q_{nuc} is 1, assuming that all nuclei have singly-degenerate nuclear states. Similarly, the minimum value of q_{rot} for a molecule is 1, which would be at absolute zero (and thus the molecule would be in the $J = 0$ rotational state for all possible rotations). At any temperature above $T = 0$ K, the q_{rot} would be greater than 1. The minimum value of q_{vib} is zero (see the examples in exercise 18.13) if either T were very low (i.e. 0 K) or the vibrational frequencies were very large. However, for any nonzero temperature and finite vibrational frequency, q_{vib} is greater than zero.

18.17. Using the expanded form of equation 18.34 (in which the definition of θ_r is given explicitly): $\dfrac{q_{rot}(H_2)}{q_{rot}(D_2)} = \dfrac{8\pi^2 I_{H_2}kT/2h^2}{8\pi^2 I_{D_2}kT/2h^2} = \dfrac{I_{H_2}}{I_{D_2}} = \dfrac{\mu_{H_2}r^2}{\mu_{D_2}r^2} = \dfrac{\mu_{H_2}}{\mu_{D_2}} = \dfrac{1}{2}$.

18.19. Acetylene with one deuterium substituted for a hydrogen should not show the same intensity variations in its rovibrational spectrum. Because the molecule no longer has a center of symmetry, the symmetry implications on the allowed rotational states no longer apply.

18.21. For NH_3: ammonia is a symmetric top, so we use equation 18.41 and the three rotational temperatures (two of them the same) from Table 18.3:

$q_{rot} = \dfrac{\pi^{1/2}}{3}\left(\dfrac{298\,K}{13.6\,K}\right)\left(\dfrac{298\,K}{8.92\,K}\right)^{1/2} = 74.8$

Carbon tetrachloride, CCl_4, is a spherical top, so we can use equation 18.38:

$q_{rot} = \dfrac{\pi^{1/2}}{12}\left(\dfrac{298\,K}{0.0823\,K}\right)^{3/2} = 32182$

18.23. C_p will always be greater than C_v, due to the additional Nk term in it. Note that this is consistent with the conclusion from phenomenological thermodynamics, which states that C_p and C_v differ by R.

18.25. Table 18.5 gives a useful summary of the contributions of each partition function to the thermodynamic property. Let us simply use those terms for a diatomic molecule, looking up the values that we need from the various tables, and solve.
For E:

$E = \dfrac{3}{2}RT - D_e + RT\left(\dfrac{\theta_v}{2T} + \dfrac{\theta_v/T}{e^{\theta_v/T}-1}\right) + RT$

$$E = \frac{3}{2}(8.314 \text{ J/mol} \cdot \text{K})(298 \text{ K}) - 431.6 \text{ kJ/mol} + (8.314 \text{ J/mol} \cdot \text{K})(298 \text{ K})\left(\frac{4227 \text{ K}}{2 \cdot 298 \text{ K}}\right.$$

$$\left. + \frac{4227 \text{ K}/298 \text{ K}}{e^{4227\text{K}/298\text{K}} - 1}\right) + (8.314 \text{ J/mol} \cdot \text{K})(298 \text{ K})$$

$E = 3.72 \text{ kJ/mol} - 431.6 \text{ kJ/mol} + 17.57 \text{ kJ/mol} + 2.48 \text{ kJ/mol} = -407.8 \text{ kJ/mol}$

For H:

$$H = \frac{5}{2}RT - D_e + RT + RT\left(\frac{\theta_v}{2T} + \frac{\theta_v/T}{e^{\theta_v/T} - 1} + 1\right) + 2RT$$

$$H = \frac{7}{2}(8.314 \text{ J/mol} \cdot \text{K})(298 \text{ K}) - 431.6 \text{ kJ/mol} + (8.314 \text{ J/mol} \cdot \text{K})(298 \text{ K})\left(\frac{4227 \text{ K}}{2 \cdot 298 \text{ K}}\right.$$

$$\left. + \frac{4227 \text{ K}/298 \text{ K}}{e^{4227\text{K}/298\text{K}} - 1} + 1\right) + 2(8.314 \text{ J/mol} \cdot \text{K})(298 \text{ K})$$

$E = 8.67 \text{ kJ/mol} - 431.6 \text{ kJ/mol} + 20.05 \text{ kJ/mol} + 4.96 \text{ kJ/mol} = -397.9 \text{ kJ/mol}$

For S, the ground electronic state is a singlet, so the electronic participation is zero:

$$S = R\left[\ln\left\{\left(\frac{2\pi mkT}{h^2}\right)^{3/2}\frac{Ve^{5/2}}{N}\right\}\right] + R\left[\frac{\theta_v/T}{e^{\theta_v/T} - 1} - \ln\left(1 - e^{-\theta_v/T}\right)\right] + R\ln\frac{T}{\theta_r} + R$$

$$S = (8.314 \text{ J/mol} \cdot \text{K})\left[\ln\left(\frac{2\pi(6.06\times10^{-26} \text{ kg})(1.381\times10^{-23} \text{ J/K})(298 \text{ K})}{(6.626\times10^{-34} \text{ J} \cdot \text{s})^2}\right)^{3/2}\frac{0.02445\text{m}^3 e^{5/2}}{6.02\times10^{23}}\right]$$

$$+ (8.314 \text{ J/mol} \cdot \text{K})\left[\frac{4227 \text{ K}/298 \text{ K}}{e^{4227/298} - 1} - \ln\left(1 - e^{-4227/298}\right)\right] + (8.314 \text{ J/mol} \cdot \text{K})\ln\frac{298 \text{ K}}{15.2 \text{ K}} + 8.314 \text{ J/mol} \cdot \text{K}$$

$S = 153.59 \text{ J/mol} \cdot \text{K} + 8.73\times10^{-5} \text{ J/mol} \cdot \text{K} + 24.74 \text{ J/mol} \cdot \text{K} + 8.314 \text{ J/mol} \cdot \text{K} = 186.6 \text{ J/mol} \cdot \text{K}$

Finally, for G, it is obvious from Table 18.5 that $G = H - TS$; so,
$G = -397.9 - (298)(0.1866) \text{ kJ/mol} = -453.5 \text{ kJ/mol}$
(but feel free to use the expressions in Table 18.5 to determine G; you should get the same answer, within truncation error).

18.27. From Tables 18.1 and 18.3, we get θ_v and θ_r for H_2: 6215 and 85.4 K, respectively. By calculating the reduced masses of H_2, HD, and D_2, we can determine the following relationships between them: $\mu(H_2)/\mu(HD) = 0.5/0.66666... = 0.75$ and $\mu(H_2)/\mu(D_2) = 0.5/1 = 0.5$. From these ratios and knowing the relationships between the θ_v and θ_r and μ, we get 5382 K and 64.0 K for θ_v and θ_r of HD (respectively), and 4395 K and 42.7 for θ_v and θ_r of D_2 (respectively).

In terms of the changes in H and S, the only contributions will be from rotational and vibrational partition functions, since the translational and electronic contributions are the same, while nuclear contributions are ignored. In addition, for ΔH, the rotational contributions are the same as well, leaving only changes in vibrations. Thus, we have for ΔH:

$$\Delta H = 2(8.314 \text{ J/mol} \cdot \text{K})(298 \text{ K})\left(\frac{5382 \text{ K}}{2(298 \text{ K})} + \frac{5382 \text{ K}/298 \text{ K}}{e^{5382 \text{ K}/298 \text{ K}} - 1} + 1 \right)$$

$$- \left[(8.314 \text{ J/mol} \cdot \text{K})(298 \text{ K})\left(\frac{6215 \text{ K}}{2(298 \text{ K})} + \frac{6215 \text{ K}/298 \text{ K}}{e^{6215 \text{ K}/298 \text{ K}} - 1} + 1 \right) \right.$$

$$\left. + (8.314 \text{ J/mol} \cdot \text{K})(298 \text{ K})\left(\frac{4398 \text{ K}}{2(298 \text{ K})} + \frac{4398 \text{ K}/298 \text{ K}}{e^{4398 \text{ K}/298 \text{ K}} - 1} + 1 \right) \right]$$

$\Delta H = 49701 - 28313 - 20760 = 628 \text{ J/mol} = 0.628 \text{ kJ/mol}$

$$\Delta S = 2 \cdot (8.314 \text{ J/mol} \cdot \text{K})\left(\frac{5382 \text{ K}/298 \text{ K}}{e^{5382 \text{ K}/298 \text{ K}} - 1} - \ln\left(1 - e^{-5382 \text{ K}/298 \text{ K}}\right) \right) + 2 \cdot (8.314 \text{ J/mol} \cdot \text{K})\left(\ln\frac{298 \text{ K}}{64.0 \text{ K}} + 1 \right)$$

$$- \left[(8.314 \text{ J/mol} \cdot \text{K})\left(\frac{6215 \text{ K}/298 \text{ K}}{e^{6215 \text{ K}/298 \text{ K}} - 1} - \ln\left(1 - e^{-6215 \text{ K}/298 \text{ K}}\right) \right) + (8.314 \text{ J/mol} \cdot \text{K})\left(\ln\frac{298 \text{ K}}{85.4 \text{ K}} + 1 \right) \right.$$

$$\left. + (8.314 \text{ J/mol} \cdot \text{K})\left(\frac{4395 \text{ K}/298 \text{ K}}{e^{4395 \text{ K}/298 \text{ K}} - 1} - \ln\left(1 - e^{-4395 \text{ K}/298 \text{ K}}\right) \right) + (8.314 \text{ J/mol} \cdot \text{K})\left(\ln\frac{298 \text{ K}}{42.7 \text{ K}} + 1 \right) \right]$$

$\Delta S = 42.21 - 18.70 - 24.47 \text{ J/mol} \cdot \text{K} = -0.96 \text{ J/mol} \cdot \text{K}$

18.29. Using the expression $E = kT^2 \left(\frac{\partial \ln q^N}{\partial T} \right) = NkT^2 \frac{1}{q} \frac{\partial q}{\partial T}$: the expression for the

translational energy is derived in chapter 17 (see equation 17.56 and the expressions leading up to it), so won't be repeated here. For the electronic contribution to the energy, we have:

$$E_{\text{elect}} = NkT^2 \frac{1}{g_1 e^{D_e/kT}} \frac{\partial \left(g_1 e^{D_e/kT} \right)}{\partial T} = NkT^2 \frac{1}{g_1 e^{D_e/kT}} \cdot g_1 \cdot -\frac{D_e}{kT^2} e^{D_e/kT}$$

The degeneracies cancel, the exponentials cancel, the kT^2 terms cancel. The only terms that don't cancel are N and $-D_e$, giving us $E_{\text{elect}} = -ND_e$. For (polyatomic) rotations, we have

$$E_{\text{rot}} = NkT^2 \frac{1}{\frac{\pi^{1/2}}{\sigma}\left(\frac{T^3}{\theta_A \theta_B \theta_C} \right)^{1/2}} \frac{\partial \left(\frac{\pi^{1/2}}{\sigma}\left(\frac{T^3}{\theta_A \theta_B \theta_C} \right)^{1/2} \right)}{\partial T}$$

$$= NkT^2 \frac{1}{\frac{\pi^{1/2}}{\sigma}\left(\frac{T^3}{\theta_A \theta_B \theta_C} \right)^{1/2}} \cdot \frac{\pi^{1/2}}{\sigma}\left(\frac{1}{\theta_A \theta_B \theta_C} \right)^{1/2} \cdot \frac{3}{2} T^{1/2}$$

All of the terms (π, σ, θs) cancel, and what we have left is (after collecting all of our terms in T):

$$E = \frac{3}{2} NkT^2 \frac{T^{1/2}}{T^{3/2}} = \frac{3}{2} NkT^2 \frac{1}{T} = \frac{3}{2} NkT$$

18.31. If the reaction deals only with isotopes, the translational and electronic partition functions cancel, and we will assume that the nuclear partition function doesn't affect the equilibrium (which may not be the case, since different nuclei are involved. However, we will make the assumption). Thus, the only contributions to the equilibrium constant come from the vibrational and rotational partition functions. For the given reaction, we would estimate K as

$$K = \frac{\left(q_{vib}q_{rot}\right)_{mixed}^2}{\left(q_{vib}q_{rot}\right)_{^{14}N_2}\left(q_{vib}q_{rot}\right)_{^{15}N_2}}.$$

Using the high-temperature limit expressions for each q, this becomes

$$K = \frac{\left(\dfrac{T}{\theta_r}\cdot\dfrac{T}{\theta_v}\right)_{mixed}^2}{\left(\dfrac{T}{2\theta_r}\cdot\dfrac{T}{\theta_v}\right)_{^{14}N_2}\left(\dfrac{T}{2\theta_r}\cdot\dfrac{T}{\theta_v}\right)_{^{15}N_2}}$$

The T variables cancel. On the assumption that the θ_v and θ_r values are similar, they too would cancel and all that would remain is

$$K = \frac{1}{(1/2)^2} = 4$$

Thus, isotope exchange reactions are predicted to have a limiting value of K of 4.

18.33. It might be tempting to think that the values for H and G are reasonably close to the actual values of these energies, much like the statistically-determined entropy S values are very close to the experimentally-determined values. However, one part of the molecular partition function is based on an arbitrary energy value: the electronic partition function, which is defined in terms of the dissociation energy D_e, which is itself defined in terms of an arbitrary zero point. Note that according to Table 18.5, S does not have that problem. Finally, for atomic systems, the total energies are equal to what is predicted from the kinetic theory of gases and does not include any contributions from electronic or nuclear energies, so H or G cannot be construed as the "total energy" of the atoms.

CHAPTER 19. THE KINETIC THEORY OF GASES

19.1. A postulate is a statement that is assumed but not proven. Other statements can be based on them and supported or not supported by them, but the postulate itself is only supported based on what is derived from it.

19.3. The drawing is left to the student.

19.5. Let us rewrite equation 19.8 in such a way as to get the expression for kinetic energy all by itself on one side. We get $\frac{1}{2}mv^2 = \frac{3pV}{2N}$. Note that the right side of this expression has pressure, volume, and amount – three of the four variables that the conditions of a gas depend on. The fourth variable is temperature, and by comparing the right side to the ideal gas law, we can argue that $\frac{3pV}{2N}$ is proportional to temperature, T. (We just don't know the proportionality constant yet.) Therefore, we substitute: $\frac{1}{2}mv^2 \propto T$, leading to the conclusion that the (average) kinetic energy of a gas is dictated solely by the temperature of the gas.

19.7. Cesium atoms have a mass of 0.1329 kg/mole:

$$200\,\text{m/s} = \sqrt{\frac{3 \cdot (8.314\,\text{J/mol}\cdot\text{K})T}{0.1329\,\text{kg/mol}}} \qquad T = 213\,\text{K}$$

$$400\,\text{m/s} = \sqrt{\frac{3 \cdot (8.314\,\text{J/mol}\cdot\text{K})T}{0.1329\,\text{kg/mol}}} \qquad T = 853\,\text{K}$$

$$600\,\text{m/s} = \sqrt{\frac{3 \cdot (8.314\,\text{J/mol}\cdot\text{K})T}{0.1329\,\text{kg/mol}}} \qquad T = 1920\,\text{K}$$

$$800\,\text{m/s} = \sqrt{\frac{3 \cdot (8.314\,\text{J/mol}\cdot\text{K})T}{0.1329\,\text{kg/mol}}} \qquad T = 3410\,\text{K}$$

$$1000\,\text{m/s} = \sqrt{\frac{3 \cdot (8.314\,\text{J/mol}\cdot\text{K})T}{0.1329\,\text{kg/mol}}} \qquad T = 5330\,\text{K}$$

The temperature is related to the square of the speed, rather than directly to the speed.

19.9. We need to solve the integral $\int_{-\infty}^{+\infty} Ae^{(1/2)Kv^2}\,dv = 1$ for the proper value of A (this is equivalent to normalizing a wavefunction…). This is an even function centered about zero, so we can rewrite this integral as $2\int_{0}^{+\infty} Ae^{(1/2)Kv^2}\,dv = 1$. The integral is of the form $\int_{0}^{+\infty} e^{-bx^2}\,dx = \frac{1}{2}\left(\frac{\pi}{b}\right)^{1/2}$. In our case, the constant $b = -\frac{1}{2}K$. Therefore, we have:

$$2A \int_0^\infty e^{(1/2)Kv^2}\, dv = 1 \qquad 2A\left[\frac{1}{2}\left(\frac{\pi}{-K/2}\right)^{1/2}\right] = 1 \qquad A\left(-\frac{2\pi}{K}\right)^{1/2} = 1 \qquad A = \left(-\frac{K}{2\pi}\right)^{1/2}$$

This final expression is equation 19.25.

19.11. Showing that K is also equal to $\dfrac{1}{v_y}\dfrac{g_y{}'(v_y)}{g_y(v_y)}$ and $\dfrac{1}{v_z}\dfrac{g_z{}'(v_z)}{g_z(v_z)}$ follows the same steps as presented in the text, starting with equation 19.16 and ending with equation 19.21. The only difference is that the derivatives in the steps are taken with respect to v_y or v_z instead of v_x. Rather than repeat the steps, the reader is referred to that section of the text.

19.13. $\dfrac{v_{rms}}{v_{most\ prob}} = \dfrac{\sqrt{3RT/M}}{\sqrt{2RT/M}} = \sqrt{\dfrac{3}{2}} = 1.2247\ldots$

19.15. Rubidium atoms have a mass of 85.5 g/mol or 0.0855 kg/mol. Using the formula for the most probable velocity:

$$0.01\,\text{m/s} = \sqrt{\frac{2(8.314\,\text{J/mol}\cdot\text{K})T}{0.0855\,\text{kg/mol}}} \qquad T = 5.14\times10^{-7}\,\text{K}$$

19.17. The relative values of the three different average velocities will always be the same. In fact, their ratios are also invariant (as shown by exercise 19.13).

19.19. Equation 19.41 is derived explicitly in the text, starting with equation 19.40. Rather than repeat it here, please refer to that section of the text.

19.21. Although argon is an atom, hydrogen is a molecule. The fact that there are two atoms of hydrogen bonded together gives H_2 an effective molecular diameter commensurate with those of other atoms.

19.23. The average collision frequency depends on the density (which relates how many gas particles there are per volume) and the temperature (which relates how fast the has particles are moving).

19.25. $1\,\text{s}^{-1} = \dfrac{\pi\rho(4.0\times10^{-10}\,\text{m})^2\left[16(1.381\times10^{-23}\,\text{J/K})(298\,\text{K})\right]^{1/2}}{(\pi\cdot2.18\times10^{-25}\,\text{kg})^{1/2}}$

$\rho = 6.42\times10^{15}/\text{m}^3$. That is, there need to be 6.42×10^{15} atoms of Xe per every cubic meter. Since 1 mole of Xe has 6.02×10^{23} atoms, we need $\dfrac{6.02\times10^{23}}{6.42\times10^{15}/\text{m}^3} = 9.38\times10^7\,\text{m}^3 = 9.38\times10^{10}\,\text{L}$.

That's the volume of a cube that is 454 meters (just under half a kilometer) on a side.

19.27. The total number of collisions is $Z\cdot V = 3.21\times10^{15}\,\text{s}^{-1}\text{m}^{-3}\times9.38\times10^7\,\text{m}^3 = 3.01\times10^{23}$ collisions per second.

19.29. What the question is asking is which quantity is larger, $z_{He \to Ar}$ or $z_{Ar \to He}$? The value of z depends on the density of the other gas. In this case, since the two gases have equal concentrations, we expect that the two values of z should be the same.

19.31. Using equation 19.51:

$$rate = 0.10 \text{ mm}^2 \times \frac{(1 \text{ m})^2}{(1000 \text{ mm})^2} \times 0.0014 \text{ mmHg} \times \frac{1 \text{ atm}}{760 \text{ mmHg}} \times \frac{101{,}325 \text{ Pa}}{1 \text{ atm}}$$

$$\times \left(\frac{1}{2\pi (0.2006 \text{ kg/} 6.02 \times 10^{23})(1.381 \times 10^{-23} \text{ J/K})(295 \text{ K})} \right)^{1/2}$$

$rate = 2.02 \times 10^{14} \text{ s}^{-1}$

That is, 2.02×10^{14} mercury atoms are diffusing through per second.

19.33. Generally, we can use equation 19.51 for this, but again we need to watch our units. Pressure should be expressed in Pa units, so we convert the 0.100 torr:

$$0.100 \text{ torr} \times \frac{1 \text{ atm}}{760 \text{ torr}} \times \frac{101325 \text{ Pa}}{1 \text{ atm}} = 13.33 \text{ Pa}$$

If the diameter of the tube is 0.01625 inches, its radius is 0.008125 inches, or 0.206375 mm. Thus, its area is $\pi r^2 = \pi (0.206375 \text{ mm})^2 = 0.1338 \text{ mm}^2 = 1.338 \times 10^{-7} \text{ m}^2$. Now substituting into equation 19.51:

$$rate = (1.338 \times 10^{-7} \text{ m}^2)(13.33 \text{ N/m}^2) \left(\frac{1}{2\pi (.0399 \text{ kg/} 6.02 \times 10^{23})(1.381 \times 10^{-23} \text{ J/K})(300 \text{ K})} \right)^{1/2}$$

$$rate = 4.294 \times 10^{16} \text{ s}^{-1} \times \frac{1 \text{ mol}}{6.02 \times 10^{23}} \times \frac{39.9}{\text{mol}} = 2.85 \times 10^{-6} \text{ g/s} = 2.85 \text{ micrograms per second}$$

19.35. Substituting only units into the expression for D_{12} in equation 19.54:

$$\sqrt{\frac{(\text{J/mol} \cdot \text{K})(\text{K})}{\text{kg/mol}}} \cdot \frac{1}{\text{m}^2 \cdot 1/\text{m}^3}$$

The other quantities in equation 19.54 don't have units, and recall that ρ is the particle density of the gases, meaning that it's the number of particles per cubic meter. The Kelvin and mole units in the square root term cancel, and we can decompose the joule unit to cancel the kilogram unit. What's left is

$$\sqrt{\frac{\text{m}^2}{\text{s}^2}} \cdot \frac{1}{\text{m}^2 \cdot 1/\text{m}^3} = \frac{\text{m}}{\text{s}} \cdot \frac{1}{1/\text{m}} = \frac{\text{m}}{\text{s}} \cdot \text{m} = \frac{\text{m}^2}{\text{s}}$$

as the standard unit on D_{12}. As mentioned in the text, D_{12} values are more often expressed in cm^2/s units.

19.37. Diffusion stops because the term $\dfrac{dc_1}{dx}$, the concentration gradient, is zero when a minor component is evenly distributed throughout the mixture.

19.39. If $E = 3/2 \, kT$, then the energy of an ammonia molecule at 295 K is

$E = 3/2 \ (1.381 \times 10^{-23} \ \text{J/K})(295 \ \text{K}) = 6.11 \times 10^{-21} \ \text{J}$

If this energy were kinetic energy, then we can use the equation $\frac{1}{2} \ mv^2$ to determine the speed of the molecule:

$$6.11 \times 10^{-21} \ \text{J} = \frac{1}{2} \left(\frac{0.017 \ \text{kg}}{6.02 \times 10^{23}} \right) v^2 \qquad v = 658 \ \text{m/s}$$

Let us use the definition of root-mean-square speed to determine another velocity for ammonia molecules using a different route:

$$v_{\text{rms}} = \sqrt{\frac{3RT}{M}} = \sqrt{\frac{3(8.314 \ \text{J/mol} \cdot \text{K})(295 \ \text{K})}{0.017 \ \text{kg/mol}}} = 658 \ \text{m/s}$$

This coincidence shouldn't be surprising, as the two expressions for speed are related via derivation of the root-mean-square speed.

19.41. (a) If the ammonia were diffusing through helium, diffusion will likely be faster. That's because the value of $(r_1 + r_2)^2$ will likely be smaller (since helium atoms are smaller than air molecules), and the reduced mass μ of NH_3-He is smaller as well. Both of these terms are in the denominator of the expression for D_{12} in equation 19.54, so D_{12} itself will be larger, implying faster diffusion. (b) If ammonia were diffusing through SF_6, the arguments are reversed. The value of $(r_1 + r_2)^2$ will be larger, and so will μ. Therefore, D_{12} will be smaller and diffusion slower.

19.43. It may be naïve to assume that ions in solution behave like individual gas particles, especially since they are charged particles. Furthermore, ions don't travel as individual particles, as the kinetic theory assumes for gas particles. They move with groups of water molecules or other ligands surrounding them, suggesting that the assumption that the travel of ions in solution should not be based on the masses of the ions themselves.

CHAPTER 20. KINETICS

20.1. For the reaction $aA + bB \rightarrow cC + dD$, we can write several other forms of the rate:

$$\text{rate} = -\frac{d[A]}{dt} = +\frac{a}{c}\frac{d[C]}{dt} \qquad \text{rate} = -\frac{d[A]}{dt} = +\frac{a}{d}\frac{d[D]}{dt}$$

or, in terms of the concentration of B:

$$\text{rate} = -\frac{d[B]}{dt} = +\frac{b}{c}\frac{d[C]}{dt} \qquad \text{rate} = -\frac{d[B]}{dt} = +\frac{b}{d}\frac{d[D]}{dt}$$

There are several other possibilities in terms of $d[C]/dt$ and $d[D]/dt$ as well.

20.3. If 1.00 mmol of H^+ are consumed, then there are $\frac{5}{16}(1.00\,\text{mmol}) = 0.3125$ mmol of Fe (s) is consumed. Therefore, the rate with respect to Fe (s) is

$$\frac{1}{5}\frac{0.3125 \times 10^{-3}\,\text{mol Fe}}{153.8\,\text{s}} = 4.06 \times 10^{-7}\,\text{mol/s}$$ (which is the same value it was with respect to H^+ in exercise 20.3). If 1.00 mmol of H^+ are consumed, then there are $\frac{2}{16}(1.00\,\text{mmol}) = 0.125$ mmol of MnO_4^- is consumed. Therefore, the rate with respect to MnO_4^- is

$$\frac{1}{2}\frac{0.125 \times 10^{-3}\,\text{mol Fe}}{153.8\,\text{s}} = 4.06 \times 10^{-7}\,\text{mol/s}$$ (which is the same value it was with respect to H^+ in exercise 20.3). In fact, all of the rates for each species turn out to be 4.06×10^{-7} mol/s.

20.5. The species could be a catalyst, or perhaps an inert gas or solvent that provides collisional energy to the reactants or products.

20.7. First, let us convert the times into rates by dividing the number of moles of A reacted by the seconds elapsed. We get

$$\frac{0.10\,\text{mol}}{36.8\,\text{s}} = 2.72 \times 10^{-3}\,\text{M/s} \qquad \frac{0.10\,\text{mol}}{25.0\,\text{s}} = 4.00 \times 10^{-3}\,\text{M/s} \qquad \frac{0.10\,\text{mol}}{10.0\,\text{s}} = 1.00 \times 10^{-2}\,\text{M/s}$$

respectively, for the three trials. Using the first and second trials:

$$\frac{2.72 \times 10^{-3}\,\text{M/s}}{4.00 \times 10^{-3}\,\text{M/s}} = \frac{k(0.20)^a (0.40)^b}{k(0.20)^a (0.60)^b}$$, which reduces to $0.68 = 0.667^a$. By inspection, it should be clear that $a = 1$. Using the second and third trials:

$$\frac{4.00 \times 10^{-3}\,\text{M/s}}{1.00 \times 10^{-2}\,\text{M/s}} = \frac{k(0.20)^a (0.60)^b}{k(0.50)^a (0.60)^b}$$, which reduces to $0.4 = 0.4^a$. By inspection, it should be clear that $b = 1$ also. Using the first set of data to determine k:

$$2.72 \times 10^{-3}\,\text{M/s} = k(0.20\,\text{M})^1 (0.40\,\text{M})^1 \qquad k = 3.4 \times 10^{-2}\,\text{M}^{-1}\text{s}^{-1}$$

20.9. If a rate is given in units of M/s, in order for that to be the ultimate unit, k would have to have units of $M^{-3}s^{-1}$.

20.11. Equation 20.15 is not written in the form of a straight line, despite the fact that the equation has the y variable on the left side and the x variable on the right side. A straight-line equation has the form $y = mx + b$, whereas this equation is of the form $y = e^x$, which would not plot as a straight line.

20.13. First, we need to determine how many grams of HgO needs to decompose in order to make 1.00 and 10.0 mL of O_2 gas at STP. Using the fact that at STP there are 22.4 L of gas volume:

$$1.00 \text{ mL} \times \frac{1 \text{ L}}{1000 \text{ mL}} \times \frac{1 \text{ mol}}{22.4 \text{ L}} \times \frac{2 \text{ mol HgO}}{1 \text{ mol O}_2} \times \frac{216.6 \text{ g HgO}}{\text{mol HgO}} = 0.0193 \text{ g HgO}$$

must decompose in order to form 1.00 mL of O_2. Of course, for 10.0 mL, ten times that amount, or 0.193 g HgO, need to be decomposed. If we start with 1.00 g, then there are $1.00 - 0.0193 = 0.9807$ grams left over $(= [A]_t)$ after 1.00 mL of oxygen are made, and $1.00 - 0.193 = 0.807$ g of HgO left over $[= [A]_t]$ after 10.0 mL of oxygen are made. Using equation 20.15:

(a) $0.9807 \text{ g} = 1.00 \text{ g} \times e^{-(6.02 \times 10^{-4} \text{ s}^{-1})t}$ Taking the logarithm of both sides:

$\ln 0.9807 = \ln 1.00 - (6.02 \times 10^{-4} \text{ s}^{-1})t$ $t = 32.4 \text{ s}$

(b) $0.807 \text{ g} = 1.00 \text{ g} \times e^{-(6.02 \times 10^{-4} \text{ s}^{-1})t}$ Taking the logarithm of both sides:

$\ln 0.807 = \ln 1.00 - (6.02 \times 10^{-4} \text{ s}^{-1})t$ $t = 356 \text{ s}$

20.15. Start with equation 20.19: $-\dfrac{d[A]}{[A]^2} = k \cdot dt$. Integrate both sides, with the left side's limits being $[A]_0$ through $[A]_t$ and the right side's limits are 0 to some time t:

$$\int_{[A]_0}^{[A]_t} -\frac{d[A]}{[A]^2} = \int_0^t k \cdot dt$$

The integral on the left side is simple $1/[A]$, evaluated at the limits, while the integral on the right is simply kt, integrated at its limits. We have:

$$\frac{1}{[A]}\Bigg|_{[A]_0}^{[A]_t} = kt\Big|_0^t$$

Evaluating at the limits, we get

$$\frac{1}{[A]_t} - \frac{1}{[A]_0} = kt$$

(which is equation 20.20).

20.17. (a) The rate law is $-\dfrac{d[A]}{dt} = k[A]^3$. We can rearrange this to get $-\dfrac{d[A]}{[A]^3} = k \cdot dt$, which ultimately integrates to $\dfrac{1}{2[A]_t^2} - \dfrac{1}{2[A]_0^2} = kt$.

(b) If we rearrange the integrated expression to $\dfrac{1}{2[A]_t^2} = kt + \dfrac{1}{2[A]_0^2}$, we see that we would have to plot $1/2[A]_t^2$ versus time t to get a straight line having slope k and intercept $1/2[A]_0^2$.

20.19. For a zeroth order reaction, the integrated rate law is written as $[A]_0 - [A]_t = kt$, which can be rewritten as $[A]_t = -kt + [A]_0$. This shows that if $[A]_t$ were plotted versus time, the slope would be $-k$ and the intercept would be $[A]_0$.

20.21. For a zeroth-order reaction, $[A]_0 - [A]_t = kt$, so we want to know the value of t when $[A]_t = 0$:

$$[A]_0 - 0 = kt \qquad [A]_0 = kt \qquad t = [A]_0/k$$

20.23. When a reaction uses H_2O as a solvent, the concentration of H_2O is typically so relatively high that it can be considered as constant throughout the course of the reaction. This 'constant' concentration of H_2O can be incorporated into the value of the rate law constant for a pseudo first order reaction, whose rate constant will have units of s^{-1}.

20.25. Four experimentally-determined parameters can be (but are not limited to) initial amount of ethyl benzoate, initial amount of sodium hydroxide, the ratio of ethyl benzoate and sodium hydroxide, and the temperature of the system. Proper selection and understanding of such parameters is necessary when studying the kinetics of any chemical reaction.

20.27. Rate laws are typically defined for initial reaction conditions. As the reaction progresses and eventually approaches equilibrium, the original description of the kinetics for the process probably won't apply. In fact, as a reaction approaches equilibrium, its net rate decreases and eventually becomes zero. As such, it's unlikely that a zeroth-order reaction will continue at a constant rate for two complete half lives.

20.29. Equation 20.33 is $[A]_t = \dfrac{[A]_0}{(k_f + k_r)}\left(k_r + k_f e^{-(k_f + k_r)t}\right)$. If the reverse reaction is negligible, that suggests that k_r is negligible and can be ignored in the above expression. If so, equation 20.33 becomes $[A]_t = \dfrac{[A]_0}{k_f}\left(k_f e^{-k_f t}\right)$, which reduces to $[A]_t = [A]_0 e^{-k_f t}$, which is the integrated rate law for a normal first-order reaction.

20.31. The initial ratio is equal to the appropriate ratio of the rate constants, in this case k_2 divided by k_1 (because the product A-B is made by the second reaction):

$$\text{ratio} = \frac{k_2}{k_1} = \frac{3.95 \times 10^{-4}\ s^{-1}}{4.40 \times 10^{-5}\ s^{-1}} = 8.98$$

The ratio of A-B/B-C at equilibrium cannot be determined based on the information given because we don't know the rate constants for the reverse processes.

20.33. Starting with equation 20.38, $[A]_t = [A]_0 e^{-(k_1 + k_2)t}$, we take the logarithm of both sides:
$\ln [A]_t = \ln [A]_0 - (k_1 + k_2)t$
This equation is in the form $y = mx + b$, the form for a straight line. In this case, the slope would be given by the expression $-(k_1 + k_2)$. Unfortunately, the slope is the sum of the two rate constants. It is not possible from a plot such as this to determine the individual values of the two rate constants; other experiments (and/or plots) are necessary.

20.35. With the understanding that straight-line plots of equations 20.41 and 20.42 would be difficult to define and/or interpret, the best way to determine k_1 and k_2 may be by numerical trial-and-error: try to fit the experimental data by assuming various values for k_1 and k_2. Analytic solutions for the two rate constants may not be forthcoming.

20.37. At $t = 0$, the amount of ^{210}Bi will be at maximum; that will be the time before any of it decays. Mathematically, the amount of ^{206}Pb will be at maximum at $t = \infty$.

20.39. (a) When $k_1 \gg k_2$, the second rate constant is negligible with respect to the first. The equations reduce to

$$[A]_t = [A]_0 e^{-k_1 t} \quad \text{(no change)}$$

$$[B]_t = \frac{k_1 [A]_0}{-k_1}\left(e^{-k_1 t} - e^{-k_2 t}\right) = -[A]_0\left(\sim 0 - e^{-k_2 t}\right) = [A]_0\left(e^{-k_2 t}\right)$$

$$[C]_t = [A]_0\left[1 + \frac{1}{k_1}\left(k_2 e^{-k_1 t} - k_1 e^{-k_2 t}\right)\right] = [A]_0\left[1 + \frac{1}{k_1}\left(\sim 0 - k_1 e^{-k_2 t}\right)\right]$$

$$= [A]_0\left[1 + \frac{1}{k_1}\left(- k_1 e^{-k_2 t}\right)\right] = [A]_0\left[1 - e^{-k_2 t}\right]$$

(b) When $k_1 \ll k_2$, the first rate constant is negligible with respect to the second. The equations reduce to

$$[A]_t = [A]_0 e^{-k_1 t} \quad \text{(no change)}$$

$$[B]_t = \frac{k_1 [A]_0}{k_2}\left(e^{-k_1 t} - e^{-k_2 t}\right) = \frac{k_1 [A]_0}{k_2}\left(e^{-k_1 t} - \sim 0\right) = \frac{k_1 [A]_0}{k_2}\left(e^{-k_1 t}\right)$$

$$[C]_t = [A]_0\left[1 + \frac{1}{-k_2}\left(k_2 e^{-k_1 t} - k_1 e^{-k_2 t}\right)\right] = [A]_0\left[1 - \frac{1}{k_2}\left(k_2 e^{-k_1 t} - \sim 0\right)\right]$$

$$= [A]_0\left[1 - e^{-k_1 t}\right]$$

20.41. This problem needs to be answered graphically, using the techniques illustrated in example 20.8. Your final answer will depend on the accuracy of the graph, but you should get a pre-exponential factor around 4×10^{-10}.

20.43. If 295 K is the nominal base temperature, then a 10 degree increase would make the second temperature 305 K. At 295 K, if the rate constant were k, then at 305 the rate constant is $2k$. Using equation 20.52:

$$\ln \frac{k}{2k} = -\frac{E_A}{8.314 \text{ J/mol} \cdot \text{K}}\left(\frac{1}{295 \text{ K}} - \frac{1}{305 \text{ K}}\right) \qquad \text{The } k\text{'s cancel; solving for } E_A:$$

$E_A = 51,850$ J/mol $= 51.85$ kJ as the activation energy.

20.45. Using the Arrhenius equation: 9.9×10^{-12} cm^3/s $= A \cdot e^{-1900 \text{ J/mol}/(8.314 \text{ J/mol} \cdot \text{K})(1153\text{K})}$
Solving for A: $A = 1.2 \times 10^{-11}$ cm^3/s

20.47. $\ln \dfrac{k}{1.77 \times 10^{-6} /\text{M} \cdot \text{s}} = -\dfrac{20000 \text{ J/mol}}{8.314 \text{ J/mol} \cdot \text{K}}\left(\dfrac{1}{373 \text{ K}} - \dfrac{1}{298 \text{ K}}\right)$
Solve for k: $k = 8.97 \times 10^{-6}$ /M·s

20.49. (a) To determine the energy of the photon (which represents the activation energy):

$$v = \frac{c}{\lambda} = \frac{2.9979 \times 10^8 \text{ m/s}}{7.50 \times 10^{-7} \text{ m}} = 3.997 \times 10^{14} \text{ s}^{-1}$$

Therefore: $E = hv = (6.626 \times 10^{-34} \text{ J·s})(3.997 \times 10^{14} \text{ s}^{-1}) = 2.65 \times 10^{-19} \text{ J}$

On a molar basis, this is $(2.65 \times 10^{-19} \text{ J})(6.02 \times 10^{23}/\text{mol}) = 159{,}000 \text{ J/mol} = 159 \text{ kJ/mol}$

(b) $3 \times 10^{11} \text{ s}^{-1} = A e^{-159{,}000 \text{ J/mol}/(8.314 \text{ J/mol·K})(310 \text{ K})}$

Solve for A: $A = 1.8 \times 10^{38} \text{ s}^{-1}$

(c) $\ln \dfrac{k}{3 \times 10^{11}/\text{s}} = -\dfrac{159{,}000 \text{ J/mol}}{8.314 \text{ J/mol·K}} \left(\dfrac{1}{271 \text{ K}} - \dfrac{1}{310 \text{ K}} \right)$

Solve for k: $k = 4.2 \times 10^7 \text{ s}^{-1}$

20.51. (a) The first step is probably the breaking of the Br-Br bond to make two Br atom radicals. Then, a Br atom abstracts an H atom from CH_3CH_3. Then, a Br atom combines with the ethyl radical to make a product, and the cycle continues. (b) The reaction would probably go faster than chlorination, since the bromine molecule is more easily broken than Cl_2 is.

20.53. The "consecutive reaction" approach doesn't work because it doesn't implicitly recognize that the first reaction comes to equilibrium and that the concentration of B is essentially constant.

20.55. If the mechanism of the chlorination of methane presented in the text is correct and the second step is the RDS, then the rate law is rate $= k[\text{Cl}][CH_4]$. Invoking the steady-state approximation, the first step is at equilibrium: $K = \dfrac{[\text{Cl}]^2}{[\text{Cl}_2]}$. Therefore, $[\text{Cl}] = (K[\text{Cl}_2])^{1/2}$.

Substituting into the rate law: rate $= k(K[\text{Cl}_2])^{1/2}[CH_4] = k'[\text{Cl}_2]^{1/2}[CH_4]$.

20.57. If the second step is the RDS, then the rate is rate $= k[CH_4Cl]$. If the first step is in equilibrium, we can use the fact that $K = \dfrac{[CH_4Cl][\text{Cl}]}{[\text{Cl}_2][CH_4]}$ and solve for $[CH_4Cl]$:

$[CH_4Cl] = \dfrac{K[\text{Cl}_2][CH_4]}{[\text{Cl}]}$ Substitute into the original rate law:

rate $= k\left[\dfrac{K[\text{Cl}_2][CH_4]}{[\text{Cl}]} \right] = \dfrac{k'[\text{Cl}_2][CH_4]}{[\text{Cl}]}$ as the final rate law. Note that this one still has the concentration of a relatively unstable intermediate, the Cl atom.

20.59. The equivalency of the two equations is shown explicitly in the text between equations 20.64 and 20.68. Rather than repeat the discussion, readers are referred to that section of the text.

20.61. Since $V = k_2[E_0]$, let us substitute into the denominator of the second term in equation 20.68:

$\dfrac{1}{\text{rate}} = \dfrac{1}{V} + \dfrac{K}{V} \cdot \dfrac{1}{[S]}$ Multiplying the first term on the right by $[S]/[S]$:

$$\frac{1}{\text{rate}} = \frac{1}{V}\frac{[S]}{[S]} + \frac{K}{V}\cdot\frac{1}{[S]} = \frac{K+[S]}{V[S]}$$ Now we take the reciprocal of both sides to get the final

expression: $\text{rate} = \dfrac{V[S]}{K+[S]}$.

20.63. I_2 halogenation should proceed most easily since it has the weakest interhalogen bond which is most easily broken to start the chain reaction.

20.65. There are many possible reactions that can be devised, starting with initiation reactions of the formation of $H + C_2H_5$ radicals or $2\ CH_3$ radicals. There are no absolute, correct answers for a suggested mechanism.

20.67. In order of presentation, the reactions are initiation, propagation, branching, propagation, termination, and propagation.

20.69. If the first proposed Mechanism of an oscillating reaction has the second step as the rate-determining step, then the rate law is rate = $k[B][C]$. Invoking the steady-state approximation, the first step is at equilibrium: $K = \dfrac{[B]}{[A]}$. Therefore, $[B] = K[A]$. Substituting into the rate law: rate = $kK[A][C] = k'[A][C]$.

If the second proposed Mechanism of an oscillating reaction has the second step as the RDS, then the rate law is rate = $k[B][C]$. Invoking the steady-state approximation, the first step is at equilibrium: $K = \dfrac{[B]^2}{[A]}$. Therefore, $[B] = (K[A])^{1/2}$. Substituting into the rate law: rate = $k(K[A])^{1/2}[C] = k'[A]^{1/2}[C]$.

Note the slight differences in the overall rate law between the two proposed Mechanisms.

20.71. Refer to equation 20.85: the pre-exponential factor A is related to the relative entropy of the transition state. If A for the chlorine-containing reaction is lower than A for the bromine-containing reaction, equation 20.85 suggests that the proposed transition state, HCl_2, is lower in entropy with respect to the original reactants than HBr_2 is to its original reactants.

20.73. Although the expression for k in equation 20.84 contains energy and entropy terms, the energy and entropy terms relate to the transition state, not directly to the reactants or products. Therefore, we can't use equation 20.84 to justify any laws of thermodynamics. However, we can use equation 20.84 to suggest conditions that would increase the value of k, thereby suggesting a faster reaction. For the E_A term, a lower E_A (energy of activation) will increase the overall value of k. For the ΔS^* term, a higher ΔS^* would increase the overall value of k. So while "lower energy" and "higher entropy" are concepts that can be applied to equation 20.84, they do not directly relate to the laws of thermodynamics.

CHAPTER 21. THE SOLID STATE: CRYSTALS

21.1. Ionic crystals are so brittle because at the atomic level, they are held together by strong interionic forces. However, if the planes of charged particles are moved by some force, they can now face like-charged particles that would repel, breaking the crystal apart.

21.3. Unit cells can be described for polycrystalline materials only if it is understood that each unit cell may apply only to a tiny portion of the overall solid. Other than that, normal rules of unit cells apply.

21.5. This structure is also a face-centered cubic unit cell.

21.7. A proposed edge-centered cubic unit cell would actually be a simple cubic lattice, with some atoms located at tetrahedrally-oriented sites within the unit cell.

21.9. The largest atom that can fit in a body-centered cubic unit cell will have the corner atom touching the center atom, which will touch the opposite corner atom. Thus, going from one corner through the center of the unit cell to the opposite corner takes 4 radii of the atom. In terms of a, the unit cell parameter, that corner-to-corner distance is $(3)^{1/2}a$. Therefore, the maximum radius an atom in a bcc unit cell can have is $\dfrac{\sqrt{3}}{4}a$.

21.11. $1\,\text{mL} \times \dfrac{1\,\text{cm}^3}{1\,\text{mL}} \times \dfrac{(1\,\text{m})^3}{(100\,\text{cm})^3} \times \dfrac{(10^{10}\,\text{Å})^3}{(1\,\text{m})^3} = 1 \times 10^{24}\,\text{Å}^3$

21.13. To solve this problem, let us determine what mass each unit cell has, divide that mass by four, and determine what combination of iron masses (55.85 amu) and sulfur masses (32.06 amu) equal that final mass. The volume of the cubic unit cell is $(5.418\,\text{Å})^3 = 159.0\,\text{Å}^3$, and using the fact that mass = density × volume:

$$\text{mass} = (5.012\,\text{g/cm}^3) \times 159.0\,\text{Å}^3 \times \dfrac{1\,\text{cm}^3}{10^{24}\,\text{Å}^3} \times \dfrac{1\,\text{amu}}{1.6605 \times 10^{-24}\,\text{g}} = 479.9\,\text{amu}$$

This is the mass per unit cell. Dividing this by four gives 120.0 as the mass of the formula for pyrite. The closest we can get to 120.0 is if we have one iron and two sulfurs for a total mass of 119.92 amu. Therefore, we suggest that the formula for pyrite is FeS_2.

21.15. The mass of one formula unit of SiO_2 is $28.1 + 2(16.0) = 60.1$ amu. Three of these units has a mass of 180.3 amu. Now, we determine the volume of the unit cell, using equation 21.4:
volume = (4.914 A)(4.914 A)(5.405 A)
$$\times \sqrt{1 - \cos^2 90° - \cos^2 90.0° - \cos^2 120° + 2\cos 90° \cos 90.0° \cos 120°}$$
volume = $130.52\,\text{Å}^3 \times (0.75)^{1/2} = 113.03\,\text{Å}^3$
Therefore, the density is:
$$d = \dfrac{180.3\,\text{amu}}{113.03\,\text{Å}^3} \times \dfrac{10^{24}\,\text{Å}^3}{\text{cm}^3} \times \dfrac{1.6605 \times 10^{-24}\,\text{g}}{\text{amu}} = 2.65\,\text{g/cm}^3$$

The experimental value for density is 2.648 g/cm^3, very close to the calculated value.

21.17. Both hexagonal close-packed and face-centered cubic unit cells represent the most efficient use of space. One might speculate that in such crystals, interactions between individual atoms lead to an energy-minimum structure.

21.19. Using simple geometry, one can show that with respect to the line perpendicular to the crystal planes, the Bragg equation would be most simply $n\lambda = 2d\cos\theta$.

21.21. For first-order diffraction:

$\lambda = 2d\sin\theta$ $\qquad$ 1.5511 Å = 2·(5.47 Å)$\sin\theta$ $\qquad$ $\sin\theta = 0.14178$ $\qquad\qquad$ $\theta = 8.15°$

For second-order diffraction:

$2\lambda = 2d\sin\theta$ $\quad$ 2·1.5511 Å = 2·(5.47 Å)$\sin\theta$ $\qquad$ $\sin\theta = 0.28356$ $\qquad\qquad$ $\theta = 16.5°$

21.23. For a body-centered unit cell, there are 2 total atoms per cell. The volume of one unit cell is (2.8664 Å)^3 = 23.551 Å^3. If the density is 7.8748 g/cm^3, then per cubic angstrom the density is

$$7.8748\frac{g}{cm^3} \times \frac{1\,cm^3}{10^{24}\,A^3} = 7.8748 \times 10^{-24}\ g/A^3$$

If one unit cell has a volume of 23.551 Å^3, then the total mass in one unit cell is 7.8748×10^{-24} × 23.551 = 1.8546×10^{-22} g. Since there are two atoms of Fe per unit cell, then each atom has a mass of half of this, or 9.2730×10^{-23} g. Now, using the atomic mass of iron, we combine these two quantities:

$$55.9349\frac{g}{mol} \times \frac{1\,atom}{9.2730 \times 10^{-23}\,g} = 6.0320 \times 10^{23}\ atoms/mol$$

21.25. Determining the ratio of the d spacings is aided with the appropriate diagrams. Figure 21.18 shows the (111) planes, while Figures 21.19a and 21.19b show (100) and (110) planes. Let us also take advantage of Figure 21.22, which calculates the d spacing for the (111) planes. For starters, we point out that the d spacing of the (100) planes is simply the lattice parameter, whatever it is for a particular crystal. For the (110) planes, which are tilted 45° with respect to the (100) planes, we can use geometry (in a manner similar to that shown in Figure 21.22) that the d spacing for these planes is $\frac{d\sqrt{2}}{2}$, or 0.7071d. Figure 21.22 shows that the d spacing for the (111) planes is (2.91/4.11) = 0.708 of the original d spacing. So, the ratio of d spacings would be 1:0.7071:0.708. (In actuality, the (110) and (111) planes would have the same d spacing, but truncation errors have been introduced into our calculation.)

21.27. The drawing is left to the student.

21.29. The given set of Miller indices is equivalent to the (111) plane.

21.31. Following Example 21.9 in the text, we use the data to determine the d spacings for each angle and look for the proper pattern. Using Bragg's law:

$$d = \frac{\lambda}{2\sin\theta} = \frac{1.5418\,\text{A}}{2\sin15.7°} \qquad d = 2.849\,\text{A}$$

$$d = \frac{1.5418\,\text{A}}{2\sin18.2°} \qquad d = 2.468\,\text{A}$$

$$d = \frac{1.5418\,\text{A}}{2\sin26.1°} \qquad d = 1.752\,\text{A}$$

$$d = \frac{1.5418\,\text{A}}{2\sin31.1°} \qquad d = 1.492\,\text{A}$$

$$d = \frac{1.5418\,\text{A}}{2\sin32.6°} \qquad d = 1.431\,\text{A}$$

Now we list the square of their reciprocals:

$1/(2.849)^2 = 0.1232 \qquad 1/(2.468)^2 = 0.1642 \qquad 1/(1.752)^2 = 0.3258$

$1/(1.492)^2 = 0.4492 \qquad 1/(1.431)^2 = 0.4883$

The ratio of the two lowest reciprocals is $0.1232/0.1642 = 0.75$. Thus, the crystal should be face-centered cubic. That means, by consulting Table 21.3, that the first diffraction is from the (111) plane. We can use equation 21.9 to determine the unit cell parameter:

$$a = \frac{\lambda\sqrt{h^2 + k^2 + \ell^2}}{2\sin\theta} = \frac{(1.5418\,\text{A})\sqrt{1^2 + 1^2 + 1^2}}{2\sin15.7°} = 4.93\,\text{A}$$

21.33. The drawing is left to the student. It should show that successive planes diffract light of the right wavelength in a destructive-interference fashion, effectively eliminating the diffraction.

21.35. The sample may be MgO, as we would expect these isoelectronic ions to have similar scattering abilities. The other ionic compounds are composed of ions of obviously different sizes, which should be reflected in a pattern of varying intensities.

21.37. Use Table 21.4 to obtain the ionic radii of the ions involved, and take the ratio of the smaller to larger ions.

(a) For titanium sulfide: $\dfrac{r_{smaller}}{r_{larger}} = \dfrac{0.68\,\text{A}}{1.84\,\text{A}} = 0.3696$. Therefore, TiS_2 should have the rutile

structure.

(b) For barium fluoride: $\dfrac{r_{smaller}}{r_{larger}} = \dfrac{1.33\,\text{A}}{1.34\,\text{A}} = 0.9925$. Therefore, BaF_2 should have the fluorite

structrue.

(c) For potassium sulfate: $\dfrac{r_{smaller}}{r_{larger}} = \dfrac{1.33\,\text{A}}{2.30\,\text{A}} = 0.5783$. Therefore, K_2SO_4 should have the rutile

structure.

21.39. Carbon typically exists as two types of solids: a covalent-network solid (as diamond), and a planar hexagonal lattice (as graphite). In graphite, the planar sheets are far enough apart that carbon isn't close-packed.

21.41. Referring to Figure 21.29, we see that for fluorite, the coordination number for the Ca ions is 8, while the coordination number for the F ions is 4. For rutile, the coordination number for the Ti ions is 6 while the coordination for the O ions is 3. The coordination numbers are different because these aren't 1:1 ionic compounds; they are 1:2 ionic compounds.

21.43. The reactions whose energy change represent the lattice energy are:
(a) K^+ (g) + F^- (g) → KF (xtal)
(b) Mg^{2+} (g) + Se^{2-} (g) → MgSe (xtal)
(c) 2 Na^+ (g) + O^{2-} (g) → Na_2O (xtal)
(d) 2 Na^+ (g) + O_2^{2-} (g) → Na_2O_2 (xtal)

21.45. (a)

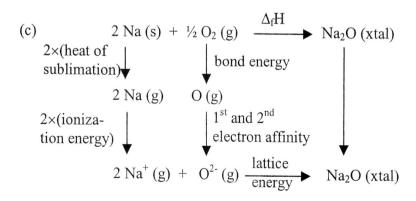

114

(d)

$$2 \text{ Na (s)} + \text{O}_2 \text{ (g)} \xrightarrow{\Delta_f H} \text{Na}_2\text{O}_2 \text{ (xtal)}$$

2×(heat of sublimation) ↓

$$2 \text{ Na (g)}$$

1st and 2nd electron affinity

2×(ionization energy) ↓

$$2 \text{ Na}^+ \text{ (g)} + \text{O}_2^{2-} \text{ (g)} \xrightarrow{\text{lattice energy}} \text{Na}_2\text{O}_2 \text{ (xtal)}$$

21.47. KI has the same crystal structure as NaCl, a face-centered cubic. We predict this using the ionic radii and determining their ratio: $\dfrac{1.33}{2.20} = 0.6045$, a ratio in the middle of the NaCl range. Therefore, we use 1.74756 as the Madelung constant. Using equation 21.13:

$$627{,}200 \text{ J/mol} = \frac{(6.02 \times 10^{23} \text{ /mol})(1.74756)(1.602 \times 10^{-19} \text{ C})^2 (1)^2}{4\pi(8.854 \times 10^{-12} \text{ C}^2 / \text{J} \cdot \text{m})(3.533 \times 10^{-10} \text{ m})} \left(1 - \frac{\rho}{3.533 \text{ A}}\right)$$

$$0.9132 = \left(1 - \frac{\rho}{3.533 \text{ A}}\right) \qquad \frac{\rho}{3.533 \text{ A}} = 0.0868 \qquad \rho = 0.307 \text{ A}$$

21.49. (a) Equation 21.13 is not applicable to solid atomic elements because they are not ions. Equation 21.13 relates the lattice energy as it relates to the force of attraction between oppositely-charged ions.
(b) 'Lattice energies' of solid atomic elements can be measured thermodynamically, by measuring how much energy it takes to melt and/or vaporize a solid element. Such measurements are not direct, however, because the interactions between uncharged species are more difficult to define than the interactions between ions.

21.51. Hydrogen is a small enough species that it would probably be incorporated as an interstitial defect in solid palladium. Other gas particles are larger than hydrogen, rationalizing why Pd absorbs hydrogen but no other gas.

21.53. For an n-type semiconductor, germanium (or any other Group IVA element) can be substituted for Ga, and selenium (or any other Group VIA element) for the As. For a p-type semiconductor, calcium (or any other Group IIA element) can be substituted for Ga, and germanium (or any other Group IVA element) can be used to substitute for As.

CHAPTER 22. SURFACES

22.1. Because there are unbalanced forces at a liquid interface, work is always performed against a force when increasing the surface area of that liquid. Work, recall, is a form of energy, implying that energy is always required to increase the surface area of a liquid.

22.3. (a) $50.0\,\text{cm}^2 \times \dfrac{27.1\,\text{dyn}}{\text{cm}} \times \dfrac{1\,\text{N}}{10^5\,\text{dyn}} \times \dfrac{1\,\text{m}}{100\,\text{cm}} = 1.355 \times 10^{-4}\,\text{N}\cdot\text{m} = 1.355 \times 10^{-4}\,\text{J}$

(b) $0.01\,\text{m}^2 \times \dfrac{27.1\,\text{dyn}}{\text{cm}} \times \dfrac{1\,\text{N}}{10^5\,\text{dyn}} \times \dfrac{100\,\text{cm}}{1\,\text{m}} = 2.71 \times 10^{-4}\,\text{N}\cdot\text{m} = 2.71 \times 10^{-4}\,\text{J}$

22.5. The surface energies of gases is so small that we can safely ignore them. Since surface energy and surface tensions are based on unbalanced forces between particles, surface tensions and surface energies of gases are miniscule under most conditions, as forces between gas particles are very small – certainly smaller than those forces in liquids and solids.

22.7. Let us use some of the results from Example 22.3. The amount of gravitational work done by the falling razor blade is 2.87×10^{-8} J, and the area involved is 7.74 cm^2. Therefore, the approximate limit to the surface tension needed to float a razor blade is

$\dfrac{2.87 \times 10^{-8}\,\text{J}}{7.74\,\text{cm}^2} \times \dfrac{10^7\,\text{erg}}{1\,\text{J}} = 0.037\,\dfrac{\text{erg}}{\text{cm}^2}$, which is a fairly low surface tension. However, we conceded that this is an estimate, and for all practical purposes the limit may be substantially higher.

22.9. (a) At first approximation, raindrops should be spherical due to surface tension. (b) Raindrops would experience a terminal velocity because the drag of the atmosphere is balancing the force due to gravity. That compensating force is acting on the bottom of the droplet, pointing up. Thus, the resulting shape of the drop would be a distorted sphere, with a flattened bottom, ultimately resembling an oblate spheroid.

22.11. You won't get the Laplace-Young equation because simply writing the area of a sphere in terms of its volume and evaluating $\delta A/\delta V$ is simply an exercise in geometry. The Laplace-Young equation was derived by realizing that the first law of thermodynamics holds for surface energies as well as any other type of energy. The derivative in terms of volume and area is a consequence of the energies involved (i.e. p-V and γ-A), not due to the shape of the system.

22.13. You would think that we would be able to minimize evaporation by increasing the external pressure. The external pressure is p_{II}, and as p_{II} increases the pressure difference decreases. However, equations 22.14 and 22.15 give expressions for Δp that are ultimately dependent only on the surface tension and the radius of the droplet, suggesting that the external pressure may not affect the Δp across the interface.

22.15. Using equation 22.14 (because we're considering a droplet, not a bubble), and converting the radii into centimeters:

(a) $\Delta p = \dfrac{2(480 \text{ dyn/cm})}{0.100 \text{ cm}} = 9600 \dfrac{\text{dyn}}{\text{cm}^2} \times \dfrac{1 \text{ N}}{10^5 \text{ dyn}} \times \dfrac{(100 \text{ cm})^2}{(1 \text{ m})^2} = 960 \text{ N/m}^2 = 960 \text{ Pa}$

This is about one-hundredth of an atmosphere.

(b) $\Delta p = \dfrac{2(480 \text{ dyn/cm})}{0.000100 \text{ cm}} = 9,600,000 \dfrac{\text{dyn}}{\text{cm}^2} \times \dfrac{1 \text{ N}}{10^5 \text{ dyn}} \times \dfrac{(100 \text{ cm})^2}{(1 \text{ m})^2} = 960,000 \text{ N/m}^2 = 960,000 \text{ Pa}$

This is about 9 and one-half atmospheres.

22.17. The drawing is left to the student.

22.19. Capillary action is not seen in cylinders with large radii because capillary action is inversely proportional to r. The greater the radius of the cylinder, the less the capillary action.

22.21. $1 \dfrac{\text{dyn}}{\text{cm}^2} \times \dfrac{(100 \text{ cm})^2}{(1 \text{ m})^2} \times \dfrac{1 \text{ N}}{10^5 \text{ dyn}} \times \dfrac{1 \text{ bar}}{10^5 \text{ Pa}} = 1 \times 10^{-6} \text{ bar}$ (where we have used the fact that 1 $\text{N/m}^2 = 1$ Pa)

22.23. Pressure times volume (essentially the left side of equation 22.21) yields energy (or energy per mole) as the ultimate unit. In order for the other side to have energy (or energy per mole) as the ultimate unit, the temperature variable must be multiplied by the conversion factor between energy and temperature. At the atomic level, this conversion factor is Boltzmann's constant, k. At the molar level, it's the ideal gas law constant R.

22.25. In a cubic crystal, atoms are closest together in one of the planes that contains two of the three axes.

22.27. The data in Table 22.2 suggest that for similar ionic compounds, as ions get smaller, the surface energy increases.

22.29. On earth, there are so many gas-phase species present that are adsorbed onto a surface that two connecting surfaces are actually separated by layers of adsorbed species. In space, without a large ambient pressure, little such gas-phase species are present, and two surfaces have the opportunity to bond more directly with each other.

22.31. One monolayer corresponds to 1 L – one-millionth of a torr of pressure exposed to a surface for one second. Thus for a pressure of 1.0×10^{-4} torr, a monolayer builds up in
$(1.0 \times 10^{-6} \text{ torr} \cdot \text{s}) = (1.0 \times 10^{-4} \text{ torr})(\text{time in seconds})$
time = 0.01 s
Thus, in about one hundredth of a second, a "clean" surface will have a monolayer of adsorbed molecules on it. It won't be clean for long!

22.33. (a) heterogeneous catalysis (b) heterogeneous catalysis (c) homogeneous catalysis (d) homogeneous catalysis (e) heterogeneous catalysis (unless the transition metal catalyst is soluble in the carbon, making an alloy; this is unlikely)

2.35. Early attempts were probably based on simple physisorption, since poly(tetrafluoroethylene) is rather chemically inert. Chemisorption would provide a better-adhering film. Today, a surface is roughened up by sandblasting and a primer is applied. The Teflon is then embedded into the primer. Thus, ultimately Teflon coatings are still physisorbed onto a surface, but the surface is better prepared so the physisorption is more durable.

22.37. If this step requires energy, it would simply add to the requirement that the exothermic steps in the mechanism must supply the energy for any endothermic processes. It is difficult to conceive that the desorption of species from a surface be exothermic (after all, energy is always required to break a chemical bond, so all bond-breaking processes are endothermic). Thus, the desorption processes should be minimally endothermic as possible to support the overall catalytic process.